Early Musical Borrowing

Criticism and Analysis of Early Music
Jessie Ann Owens, *series editor*

TONAL STRUCTURES IN EARLY MUSIC
edited by Cristle Collins Judd

COUNTERPOINT AND COMPOSITIONAL PROCESS IN THE TIME OF DUFAY
Perspectives from German Musicology
edited and translated by Kevin N. Moll

THE LANGUAGE OF THE MODES
Studies in the History of Polyphonic Modality
by Frans Wiering

COUNTERPOINT, COMPOSITION, AND *MUSICA FICTA*
by Margaret Bent

EARLY MUSICAL BORROWING
edited by Honey Meconi

Early Musical Borrowing

Edited by Honey Meconi

ROUTLEDGE
NEW YORK AND LONDON

Published in 2004 by
Routledge
29 West 35th Street
New York, NY 10001

Published in Great Britain by
Routledge
11 New Fetter Lane
London EC4P 4EE

Routledge is an imprint of the Taylor & Francis Group.

Printed in the United States of America on acid free paper.

Cover illustration: Pierre de la Rue, *Missa Ave sanctissima Maria*, Kyrie I, altus and bassus. © Brussels, Royal Library of Belgium, MS 15075, fol. 85r. Used with permission.

10 9 8 7 6 5 4 3 2 1

Cataloging-in-Publication Data is available from the Library of Congress

ISBN 0-8153-3521-0

Contents

Series Editor's Foreword

JESSIE ANN OWENS

Recent years have seen a critical reassessment of our approach to early music. Musicians, scholars, and critics have been searching for ways of talking about and reacting to the music that engage it not from the perspectives of later music but rather on its own terms. These new approaches would not be possible without the scholarship of the previous decades. The discovery and cataloguing of musical sources, the preparation of critical editions, and the investigation of archival documents have furnished important information about composers, performers, patrons, and institutions that supported the creation and performance of early music. Building on this work, the editors of these volumes now seek to develop and explore analytical methodologies for the discussion of early music as music.

Analytic methods are not easily found for early music. The theorists of the time had their own agendas, and they do not provide models that suit our purposes. As a consequence, many twentieth-century scholars have chosen approaches that reflect their own beliefs about early music and its relation to later music. While some continue to rely on common practice tonality as a prism through which to view early music, others have begun to explore methods that respect the integrity and self-sufficiency of the languages of early music.

We offer a forum for exploration of particular topics, from both methodological and critical viewpoints. Our premise is that we can best develop new methodologies by encouraging debate. We will explore compositional procedures, tonal structures, musical borrowing, and other topics, focusing both on individual compositions and on theoretical systems. We seek to encourage critical writing about music that will be useful to performers, listeners, and scholars.

Acknowledgments

This volume grew out of conversations with series editor Jessie Ann Owens, so my first thanks are to her and to Leo Balk, general editor at the time the book was conceived. Since then first Soo Mee Kwon and more recently current music editor Richard Carlin have been extraordinarily helpful and patient as the project has grown to completion. I would also like to thank Henry Bashwiner and Emily Vail of Routledge for their assistance in the process of publication. All of the authors have my deep appreciation for both their excellent essays and their ready responses to my seemingly unending requests. Murray Steib kindly provided musical examples for his own essay, while Joseph Patrick prepared all others; I gratefully acknowledge both their help and that of Rice University for its financial assistance. I must also thank Paul Ranzini, Angela Wren Wall, Andrew H. Weaver, Cathy Ann Elias, and Jenny Hodgson for contributions of various kinds. Finally, this volume would not have been possible without the support in countless direct and indirect ways of my beloved husband, Michel Godts, and our wonderful son, Yannick.

List of Abbreviations

Bibliographical Abbreviations

AIM	American Institute of Musicology
CC	*Census-Catalogue of Manuscript Sources of Polyphonic Music, 1400–1550.* 5 vols. Vol. 1 edited by Charles Hamm and Herbert Kellman. Vols. 2–5 edited by Herbert Kellman. Renaissance Manuscript Studies 1. Neuhausen-Stuttgart: Hänssler/AIM, 1979–88.
CMM	Corpus mensurabilis musicae
JAMS	*Journal of the American Musicological Society*
JosCon	Edward E. Lowinsky and Bonnie J. Blackburn, eds. *Josquin des Prez: Proceedings of the International Josquin Festival-Conference held at the Juilliard School at Lincoln Center in New York City, 21–25 June 1971.* London, New York, and Toronto: Oxford University Press, 1976.
JRMA	*Journal of the Royal Musical Association*
New Grove	*The New Grove Dictionary of Music and Musicians.* Edited by Stanley Sadie. 20 vols. London: Macmillan, 1980.
New Grove II	*The New Grove Dictionary of Music and Musicians.* 2nd ed. Edited by Stanley Sadie. 29 vols. London: Macmillan, 2001.
TVNM	*Tijdschrift van de Vereniging voor Nederlandse Muziekgeschiedenis*

Manuscript Sigla

AostaS D19	Aosta, Biblioteca del Seminario Maggiore, MS A^1 D19
BerlS 40608	Berlin, Staatsbibliothek Preussischer Kulturbesitz, MS Mus. 40608
BolC Q19	Bologna, Civico Museo Bibliografico Musicale, MS Q19
BolC Q20	Bologna, Civico Museo Bibliografico Musicale, MS Q20
BolSP 38	Bologna, Archivio Musicale della Fabbriceria di San Petronio, MS A.XXXVIII (*olim* CC)
BrusBR 215–16	Brussels, Bibliothèque Royale, MS 215–16
BrusBR 228	Brussels, Bibliothèque Royale, MS 228
BrusBR 5557	Brussels, Bibliothèque Royale, MS 5557
BrusBR 9126	Brussels, Bibliothèque Royale, MS 9126
BrusBR 15075	Brussels, Bibliothèque Royale, MS 15075
BrusBR IV.922	Brussels, Bibliothèque Royale, MS IV.922
CambraiBM 4	Cambrai, Bibliothèque Municipale, MS 4
ChiN C36	Chicago, Newberry Library, Case MS.-VM 2147.C36 (#4)
ChiN M91	Chicago, Newberry Library, Case MS.-VM 1578.M91
CoimU 2	Coimbra, Biblioteca Geral da Universidade, MS M.2
DijBM 517	Dijon, Bibliothèque Municipale, MS 517 (*olim* 295)
DresSL Pirna IV	Dresden, Sächsische Landesbibliothek, MS Pirna IV
EscSL V.III.24	Escorial, Real Monasterio de San Lorenzo del Escorial, Biblioteca y Archivo de Música, MS V.III.24
FlorC 2439	Florence, Biblioteca del Conservatorio di Musica Luigi Cherubini, MS Basevi 2439
GranCG 1	Granada, Catedral de Granada, MS 1
GuatC 4	Guatemala City, Catedral, Archivo Capitular, MS 4
's-HerAB 72A	's-Hertogenbosch, Archief van de Illustre Lieve Vrouwe Broederschap, MS 72A
's-HerAB 72B	's-Hertogenbosch, Archief van de Illustre Lieve Vrouwe Broederschap, MS 72B
JenaU 2	Jena, Universitätsbibliothek, MS 2
JenaU 3	Jena, Universitätsbibliothek, MS 3
JenaU 4	Jena, Universitätsbibliothek, MS 4
JenaU 5	Jena, Universitätsbibliothek, MS 5
JenaU 7	Jena, Universitätsbibliothek, MS 7
JenaU 9	Jena, Universitätsbibliothek, MS 9
JenaU 12	Jena, Universitätsbibliothek, MS 12
JenaU 21	Jena, Universitätsbibliothek, MS 21
JenaU 22	Jena, Universitätsbibliothek, MS 22
LonBL 31922	London, British Library, Add. Ms. 31922
LonBL 70516	London, British Library, Add. Ms. 70516

LonBLE 3307	London, British Library, MS Egerton 3307
LucAS 238	Lucca, Archivio di Stato, Biblioteca Manoscritti, MS 238
LucBC 601	Lucca, Biblioteca capitolare, MS 601
MechAS s.s.	Mechelen, Archief en Stadsbibliotheek, MS s.s.
MilA 46	Milan, Biblioteca Ambrosiana, MS E. 46. Inf.
MilD 2	Milan, Archivio della Veneranda Fabbrica del Duomo, Sezione Musicale, Librone 2 (*olim* 2268)
MilD 3	Milan, Archivio della Veneranda Fabbrica del Duomo, Sezione Musicale, Librone 3 (*olim* 2267)
ModE M.1.13	Modena, Biblioteca Estense e Universitaria, MS α.M.1.13
MontsM 750	Montserrat, Biblioteca del Monestir, MS 750
MontsM 766	Montserrat, Biblioteca del Monestir, MS 766
MontsM 773	Montserrat, Biblioteca del Monestir, MS 773
MunBS 7	Munich, Bayerische Staatsbibliothek, Musiksammlung, Musica MS 7
MunBS 26	Munich, Bayerische Staatsbibliothek, Musiksammlung, Musica MS 26
MunBS 33	Munich, Bayerische Staatsbibliothek, Musiksammlung, Musica MS 33
MunBS 37	Munich, Bayerische Staatsbibliothek, Musiksammlung, Musica MS 37
MunBS 39	Munich, Bayerische Staatsbibliothek, Musiksammlung, Musica MS 39
MunBS 47	Munich, Bayerische Staatsbibliothek, Musiksammlung, Musica MS 47
MunBS 3154	Munich, Bayerische Staatsbibliothek, Musiksammlung, Musica MS 3154
MunBS F	Munich, Bayerische Staatsbibliothek, Musiksammlung, Musica MS F
NapBN 40	Naples, Biblioteca Nazionale, MS VI E 40
ParisBN 57	Paris, Bibliothèque Nationale, Département de la Musique, MS Rés. Vmc. 57
ParisBNL 8884	Paris, Bibliothèque Nationale, Département des Manuscrits, Fonds Latin, MS 8884
PueblaC 1	Puebla (Mexico), Archivo de Música Sacra de la Catedral, MS I
RomeC 2856	Rome, Biblioteca Casanatense, MS 2856 (*olim* O.V. 208)
SensBM 6	Sens, Bibliothèque municipale, MS 6
SienBC K.I.2	Siena, Biblioteca Comunale degli Intronati, MS K.I.2
StuttL 44	Stuttgart, Württembergische Landesbibliothek, MS Musica folio I 44

StuttL 46	Stuttgart, Württembergische Landesbibliothek, MS Musica folio I 46
SubA 248	Subiaco, Monumento Nazionale dell'Abbazia di Santa Scolastica, Biblioteca Statale, MS 248
SuttonO 4	Sutton Coldfield, Oscott College, Old Library, MS Case B No. 4
ToleF 23	Toledo, Catedral, Obra y Fabrica, MS Reservado 23
TrentC 87	Trent, Museo Provinciale d'Arte, Castello del Buon Consiglio, MS 87
TrentC 88	Trent, Museo Provinciale d'Arte, Castello del Buon Consiglio, MS 88
TrentC 89	Trent, Museo Provinciale d'Arte, Castello del Buon Consiglio, MS 89
TrentC 90	Trent, Museo Provinciale d'Arte, Castello del Buon Consiglio, MS 90
TrentC 91	Trent, Museo Provinciale d'Arte, Castello del Buon Consiglio, MS 91
TrentC 92	Trent, Museo Provinciale d'Arte, Castello del Buon Consiglio, MS 92
TurBN I.27	Turin, Biblioteca Nazionale Universitaria, MS Riserva musicale I.27 (*olim* Regia Biblioteca dell'Università, q^{m} III.59)
UppsU 76b	Uppsala, Universitetsbiblioteket, MS Vokalmusik i Handskrift 76b
UppsU 76c	Uppsala, Universitetsbiblioteket, MS Vokalmusik i Handskrift 76c
VatC 234	Vatican City, Biblioteca Apostolica Vaticana, MS Chigi C.VIII.234
VatP 1976–79	Vatican City, Biblioteca Apostolica Vaticana, MSS Palatini Latini 1976–1979
VatP 1982	Vatican City, Biblioteca Apostolica Vaticana, MS Palatini Latini 1982
VatRE 2049	Vatican City, Biblioteca Apostolica Vaticana, MS Reginenses Latini 2049
VatS 5	Vatican City, Biblioteca Apostolica Vaticana, MS Cappella Sistina 5
VatS 12	Vatican City, Biblioteca Apostolica Vaticana, MS Cappella Sistina 12
VatS 14	Vatican City, Biblioteca Apostolica Vaticana, MS Cappella Sistina 14
VatS 16	Vatican City, Biblioteca Apostolica Vaticana, MS Cappella Sistina 16

VatS 34	Vatican City, Biblioteca Apostolica Vaticana, MS Cappella Sistina 34
VatS 35	Vatican City, Biblioteca Apostolica Vaticana, MS Cappella Sistina 35
VatS 36	Vatican City, Biblioteca Apostolica Vaticana, MS Cappella Sistina 36
VatS 41	Vatican City, Biblioteca Apostolica Vaticana, MS Cappella Sistina 41
VatS 45	Vatican City, Biblioteca Apostolica Vaticana, MS Cappella Sistina 45
VatS 51	Vatican City, Biblioteca Apostolica Vaticana, MS Cappella Sistina 51
VatS 55	Vatican City, Biblioteca Apostolica Vaticana, MS Cappella Sistina 55
VatS 63	Vatican City, Biblioteca Apostolica Vaticana, MS Cappella Sistina 63
VatS 160	Vatican City, Biblioteca Apostolica Vaticana, MS Cappella Sistina 160
VatSP B80	Vatican City, Biblioteca Apostolica Vaticana, MS San Pietro B 80
VenMC 1602	Venice, Museo Civico Correr, Biblioteca d'Arte e Storia Veneziana, MS Cicogna 1602
VerBC 756	Verona, Biblioteca Capitolare, MS DCCLVI
VerBC 760	Verona, Biblioteca Capitolare, MS DCCLX
VerBC 761	Verona, Biblioteca Capitolare, MS DCCLXI
VienNB 1783	Vienna, Österreichische Nationalbibliothek, Handschriften- und Inkunabelsammlung, MS 1783 (*olim* Theol. 34; VII.A.16)
VienNB 4809	Vienna, Österreichische Nationalbibliothek, Handschriften- und Inkunabelsammlung, MS 4809 (*olim* Theol. 35; VIII.A.I)
VienNB 11778	Vienna, Österreichische Nationalbibliothek, Handschriften- und Inkunabelsammlung, MS 11778 (*olim* Theol. 37; VIII.A.3)
VienNB Mus. 15495	Vienna, Österreichische Nationalbibliothek, Musiksammlung, MS Mus. 15495 (*olim* Kunsthistorisches Staatsmuseum, Sammlung für Plastik und Kunstgewerbe, MS 5248; *olim* Series nova 2660)
VienNB Mus. 15496	Vienna, Österreichische Nationalbibliothek, Musiksammlung, MS Mus. 15496 (*olim* Kunsthistorisches Staatsmuseum, Sammlung für Plastik und Kunstgewerbe, MS 5261; *olim* Series nova 2661)

VienNB Mus. 18745	Vienna, Österreichische Nationalbibliothek, Musiksammlung, MS Mus. 18745 (*olim* A.N.35.E.127)
VienNB Mus. 18832	Vienna, Österreichische Nationalbibliothek, Musiksammlung, MS Mus. 18832 (*olim* A.N.35.H.27)
WeimB A	Weimar, Bibliothek der Evangelisch-Lutherischen Kirchengemeinde, MS A
WolfA 287	Wolfenbuttel, Herzog August Bibliothek. MS Guelferbytanus 287 Extravagantium
WorcC F. 160	Worcester Cathedral, Codex F. 160

Print Sigla

1504^{1}	*Motetti C* (Venice: Petrucci, 1504)
1504^{3}	*Canti C* (Venice: Petrucci, 1504 mod. st.)
1505^{1}	*Fragmenta missarum* (Venice: Petrucci, 1505)
1509^{2}	*Frottole libro nono* (Venice: Petrucci, 1509 mod. st.)
1516^{1}	*Liber quindecim missarum...* (Rome: Antico, 1516)
1534^{5}	*Liber tertius... motetos* (Paris: Attaingnant, 1534)
1534^{9}	*Liber septimus... modulos* (Paris: Attaingnant, 1534)
1538^{9}	*Trium vocum carmina* (Nuremberg: Formschneider, 1538)
1539^{1}	*Liber quindecim missarum à praestantissimis...* (Nuremberg: J. Petreius, 1539)
1539^{10}	*Tertius liber cum quatuor vocibus. Motetti del fiore* (Lyons: Moderne, 1539)
1539^{13}	*Primus liber cum quatuor vocibus. Motetti del frutto a quatro* (Venice: Gardane, 1539)
1541^{1}	*Opus decem missarum quatuor vocum...* (Wittenberg: G. Rhaw, 1541)
1543^{16}	*Premier livre des chansons a quatre parties...* (Antwerp: Susato, 1543)
1544^{10}	*Le second livre des chansons a quatre parties...* (Antwerp: Susato, 1544)
1545^{6}	*Bicinia gallica, latina...* (Wittenberg: G. Rhaw, 1545)
1545^{14}	*Le sixiesme livre... chansons nouvelles a cincq et a six parties...* (Antwerp: Susato, 1545)
1546^{9}	*Moralis hispani... musica cum vocibus quatuor...* (Venice: A. Gardane, 1546)
1547^{5}	*Liber tertius sacrarum cantionum, quatuor vocum...* (Antwerp: Susato, 1547)
1554^{3}	*Liber tertius cantionum sacrarum... quinque et sex vocum...* (Louvain: P. Phalèse, 1554)

B4643	Antoine Brumel, *Misse* (Venice: Petrucci, 1503)
D3416	Baldassare Donato, *Il primo libro de motetti a cinque, a sei, et otto voci* (Venice: Angelo Gardano, 1599)
G58	*Concerti di Andrea, et di Gio: Gabrieli*... (Venice: Angelo Gardano, 1597)
J666	*Liber primus missarum Josquin* (Venice: Petrucci, 1502)
J670	*Missarum Josquin liber secundus* (Venice: Petrucci, 1505)
L718	Petrus de la Rue, *Misse* (Venice: Petrucci, 1503)
L795	Orlando di Lasso, *Sacrae cantiones... liber tertius* (Venice: Antonio Gardano, 1566)
M269	Pierre de Manchicourt, *Liber decimus quartus XIX musicas cantiones* (Paris: Attaingnant & Jullet, 1539)
M1270	Tiburtio Massaini, *Sacri cantus quinque paribus vocibus... liber secundus* (Venice: Angelo Gardano, 1580)
M4015	Jean Mouton, *Missarum... liber primus* (Fossombrone: Petrucci, 1515)
P664	Giovanni Pierluigi da Palestrina, *Missarum liber tertius* (Rome: Dorico, 1570)
P689	Giovanni Pierluigi da Palestrina, *Motecta festorum... liber primus* (Venice: Antonio Gardano, 1564)
P5181	Costanzo Porta, *Liber quinquaginta duorum motectorum*... (Venice: Angelo Gardano, 1580)
R3061	Vincenzo Ruffo, *Messe a cinque voci*... (Brescia: Vincenzo Sabbio, 1580)
V1006	Orazio Vecchi, *Sacrarum cantionum... liber secundus* (Venice: Angelo Gardano, 1597)
V1354	Ludovico da Viadana, *Motecta festorum... opera X* (Venice: Ricciardo Amadino, 1598)
V1421	Tomás Luis de Victoria, *Motecta, que partim quaternis*... (Venice: sons of Antonio Gardano, 1572)
W1108	Adrian Willaert, *Motetti... libro secondo a quattro voci* (Venice: Scotto, 1539)
W1110	Adrian Willaert, *Musica quinque vocum... liber primus* (Venice: Scotto, 1539)

Introduction: Borrowing and Early Music

HONEY MECONI

Borrowing is probably almost as old as music itself, and Western notated music is replete with examples from every time period.[1] Yet borrowing holds a special place in music written before 1600. Borrowed material is the foundation for much Western sacred music before that date and was the very *fons et origo* of Western polyphony itself. This collection of essays focuses on an era of especially intense interest in borrowed material, the fifteenth and sixteenth centuries (called here the Renaissance for the sake of convenience). For most sixteenth-century and many fifteenth-century masses, for example, a composer started with the choice of a model before anything else. The name of the model was sometimes proclaimed in the title of the mass (a more common occurrence the later we get in the Renaissance), while the name of the borrowed material might be given in one or more of the voice parts. At times the borrowed material was visibly different in dramatic ways from its surroundings—a short tune in long note values with many ligatures, accompanied by complex canonic instructions. Whole series of works were generated by especially famous models—*L'homme armé, Fortuna desperata, Fors seulement,* to name just a few—and an element of competition seems at play in many of these reworkings. It is the very obviousness of compositional interest in preexistent material that makes this time seem a heyday of borrowing.

The studying of borrowing has thus been a powerful tool in analysis and often the means of first approach to a composition. Borrowing is seemingly easier to talk about than text/music relations in this music (whose underlay was problematic, to say the least), modality or tonality (about the complexities of which we hardly need reminding), or other slippery elements of a Renaissance work. One found the model and traced it through the piece, simply enough—except that not much turned out to be simple once one got past the most obvious and straightforward cantus firmus use.

Major questions that have arisen are what constitutes borrowing, how we define the different types in use, what we call them, and why and how composers borrowed what they did. These topics have been the subject of fierce and impassioned debate; this is not a field for the faint of heart. Take the topic of terminology, intimately caught up with issues of definition. Here the process that has engendered the most discussion over its name is that of polyphonic borrowing. A variety of terms has been used to describe this process, but only two have gained regular currency: parody, which has a long scholarly record and a short historical pedigree, and imitation, which is much more recent in musicological circles while being older in its historical use. Although Lewis Lockwood, author of the classic article on this terminology, was at pains to describe a specific kind of compositional process employed by sixteenth-century works that actually used these terms, other twentieth-century scholars and earlier nineteenth-century writers have been less careful.[2] Many writers have seen the presence of any amount of polyphonic borrowing, whether small or large, as sufficient to label a work a parody or imitation mass (or "proto-parody" or "incipient parody" or some variant thereof), though Murray Steib's recent work has shown that virtually any mass based on a polyphonic model will use polyphonic borrowing from the primary model.[3] Scholars also take diametrically opposed positions about the kind of borrowing that constitutes parody. For some, the more literal the polyphonic borrowing the greater the amount of parody; for others, the writing must go beyond mere quotation and involve genuine reworking of the original material to qualify as parody or imitation.

Unfortunately, neither "parody" nor "imitation" has very deep terminological roots in sixteenth-century music, appearing predominantly in geographically restricted areas almost half a century after what are widely considered the earliest examples of that kind of mass writing, making neither of these words ideal for our use. A similar problem exists with our use of the term cantus firmus. Although a number of clear-cut historical uses can be identified, modern scholarship has significantly broadened its application; as Jennifer Bloxam has noted, "Cantus firmus composition is now understood to encompass a wide range of rhythmic and melodic treatments of an antecedent tune within a new polyphonic texture."[4] But the term's gain in versatility has been offset by a loss in precision: it obscures crucial large-scale differences in compositional process between pieces that use a model strictly and those that use it freely, with the result that the more fastidious prefer instead to employ the term "cantus prius factus" in describing a borrowed melody. There is also no clear agreement on the difference between a cantus firmus mass where the model is paraphrased and a so-called paraphrase mass.[5] Again, we should note the lack of any contemporary historical validity for the terminology paraphrase or paraphrase mass. This mangling of terms for borrowing is so great and our chances of concord so slim that one is tempted to agree with Saul Kripke's famous conclusion about Wittgenstein's philosophical writings, "There can be no such thing as meaning anything by any word."[6]

The issue of what is being borrowed has been similarly controversial. To make a gross generalization, one might divide borrowing into two categories, Type A (overt) and Type B (covert), with Type A being obvious and outgoing, drawing attention to itself, and Type B being hidden and more subtle. To make another gross generalization, the Renaissance might initially seem to be an era of Type A borrowing, with Type B being more indicative of, say, borrowing in the nineteenth century. But it is arguably the very prevalence of straightforward, Type A borrowing in the Renaissance that has made us less sensitive to examples of Type B borrowing, and there are plenty of pieces that borrow in ways that are not immediately apparent, such as Obrecht's structural modeling of his *L'homme armé* mass on that of Busnoys. Obviously, the farther we get from Type A borrowing the greater the potential for argument as to whether borrowing is taking place at all. Certainly scholars are happiest with clear-cut, announced, melodic borrowing, less comfortable as we explore nonmelodic borrowing, and increasingly uneasy with other, unannounced influences. Yet it is surely in these quieter types of influence that the greatest potential for discovery lies.

Even if one were to ignore the most delicate instances of musical borrowing and somehow come to agreement on terminology and definitions, would it be possible at this stage to compile a history of borrowing in the Renaissance? Almost certainly not, if for no other reason than that most basic research still remains to be done for composer after composer during this period. Except for the most famous of composers—and sometimes not even for them—it is difficult to find even thumbnail sketches of most of their works, to say nothing of detailed analyses vis-à-vis borrowing. We continue to lack complete modern editions of major figures, and perhaps most crucially, this music still remains largely silent. Little of it is in the public ear, and all too little more in our own scholarly ears. What would it do for the study of borrowing if we could turn to a vast aural repertoire, equivalent to the common knowledge of the standard practice repertoire, when we go in search of borrowings? At least some of us have been convinced of the presence or absence of a model only when we could really hear—and not just in our minds—the piece of music doing the borrowing, and then only with aural knowledge of the presumed model.

The present volume, being text, is not about to solve the problem of aural familiarity, but it does move us toward increased understanding of some aspects of borrowing. Although the essays are almost equally divided between fifteenth- and sixteenth-century music, seven of the eight concentrate on mass music, reflecting the significance that borrowing had for mass construction during this time. Only Michele Fromson ventures outside the mass, and then into still another genre of sacred music, the motet.

Three of the essays—all destined to spark lively debate—deal with the question of allusion or citation. Fromson provides the long-awaited explication of the methodology behind her 1994 article on Willaert's *Musica nova*,[7] carefully laying out the criteria for recognition of chant citation in the sixteenth-century motet. Christopher Reynolds provides a refresher course on allusion in an essay both suggesting a new interpretation of Josquin's famous *Missa Hercules dux ferrariae* and using that interpretation as evidence for placing composition of the mass in the sixteenth rather than fifteenth century. The dramatis personae of Reynolds's essay are among the most musically astute in all of Renaissance Europe, the ones most likely to recognize and comprehend musical allusions. Working with earlier music—most specifically the masses of Caron—Jenny Hodgson vigorously opposes components of allusion theory. While acknowledging that allusion exists, she argues that presumed instances of allusion in fifteenth-century masses result instead from shared compositional processes rather than conscious borrowing. Clearly, the debate about this aspect of borrowing is far from being settled.

Two studies of specific genres make very important contributions, though in completely different ways. Jennifer Bloxam addresses the vexed question of why musicians turned so often to the polyphonic chanson as model in their composition of fifteenth-century masses. Drawing on a wealth of evidence from literary, visual, and theological sources, her richly illustrated essay describes for the first time the cultural context from which this important genre grew and in which these masses must be considered to be fully understood. Andrew Weaver tackles the presumably modest genre of the polyphonic ferial mass and demonstrates unsuspected interrelationships among the various branches through the fifteenth and sixteenth centuries, providing along the way (among other things) answers to the question of priority in the case of La Rue's and Févin's ferial masses, ammunition for the debate on performance in the papal chapel, and seeds for future research on the *Missa brevis* as a genre.

Murray Steib's study of polyphonic borrowing in the masses of Johannes Ockeghem shows how knowledge of a composer's practice in this regard can provide answers to difficult questions of the authenticity or the nature of a model—a fresh approach that should win converts to this kind of analysis. In her essay, Cathy Ann Elias upsets traditional understanding of mid-sixteenth-century mass composition by demonstrating the kaleidoscopic commingling of borrowing techniques within single chanson masses of Gombert, Crecquillon, and various contemporaries, alerting future scholars to be on the lookout for different methods of borrowing. And in my own contribution, the manuscripts of the Habsburg-Burgundian court scriptorium serve as a case study for investigating the ways in which scribes acknowledged borrowing in their sources, suggesting that while the name of the model was only occasionally used in a work's formal title, the appearance of the full-fledged parody mass sparked increased awareness of borrowing in multiple voices, even if the material borrowed happened to be chant.

How and why composers borrowed, what borrowing meant to them and to their listeners, how we should approach this music and its models: all of this remains of compelling interest in the criticism and analysis of early music. These essays will provide a guide to further investigation.

Notes

1. See the extended discussion in J. Peter Burkholder, "Borrowing," *New Grove II.*
2. Lewis Lockwood, "On 'Parody' as Term and Concept in 16th-Century Music," in Jan LaRue et al., eds., *Aspects of Medieval and Renaissance Music: A Birthday Offering to Gustave Reese* (New York: W. W. Norton, 1966; reprint ed., New York: Pendragon, 1978), 560–75. For a survey of modern scholarly definitions of parody, see Cathy Ann Elias, "Imitation, Fragmentation and Assimilation of Chansons in the Masses of Gombert, Clemens, and Crecquillon: A Kaleidoscopic Process" (Ph.D. dissertation, University of Chicago, 1994), 1–26.
3. Murray Steib, "Imitation and Elaboration: The Use of Borrowed Material in Masses from the Late Fifteenth Century" (Ph.D. dissertation, University of Chicago, 1992).
4. M. Jennifer Bloxam, "Cantus firmus," *New Grove II*, v:67–74, at 67. Bloxam surveys historical uses of the term as well as discussing current understanding of the term.
5. Allan Atlas, *Renaissance Music: Music in Western Europe, 1400–1600* (New York and London: W. W. Norton, 1998), 304, offers a commonly accepted definition of paraphrase mass; a much broader use of the term appears in Gustave Reese, *Music in the Renaissance* (New York: W. W. Norton, 1954), 32 and passim. Another major history, Leeman L. Perkins's *Music in the Age of the Renaissance* (New York and London: W. W. Norton, 1999), refers to the genre only in a footnote (359n, mentioning Reese) and otherwise ignores the concept.
6. Saul A. Kripke, *Wittgenstein on Rules and Private Language: An Elementary Exposition* (Cambridge, Mass.: Harvard University Press, 1982), 55.
7. Michele Fromson, "Themes of Exile in Willaert's *Musica nova*," *JAMS* 48 (1994): 442–87.

1
A Cultural Context for the Chanson Mass[1]

M. JENNIFER BLOXAM

Sometime before ca. 1436, the composer Pierre Fontaine, chaplain at the court of Burgundy and the Sainte-Chapelle in Bourges, penned the plaintive rondeau *Mon cuer pleure*. Stock poetic images of crying hearts concealed by laughing eyes, dolorous martyrdom, and frequent sighs perfectly capture the dramatic posture of the courtly lover suffering silent anguish for want of his favorite lady's favors.[2]

Within the next twenty years, a composer named Le Rouge fashioned a mass based on Fontaine's rondeau. This new mass was probably created to adorn certain celebrations of this solemn ritual either by the chapel of the Burgundian duke Philip the Good or by that of Charles d'Orléans; both establishments numbered among their members someone by the name of Le Rouge.[3] The *Missa Mon cuer pleure* has disappeared, its existence known only because Johannes Tinctoris referred to it in his treatise *Proportionale musices* of ca. 1473–74.[4] But even in its absence this lost mass raises intriguing questions. Why did the composer select a secular love song as the basis for music destined for a religious service? What did it signify to the singers and worshipers present? Indeed, these questions merit posing for the entire and considerable body of surviving masses based on secular love songs that emerged in the 1450s and flourished well into the sixteenth century. For centuries before this decade, polyphony destined for performance within the mass liturgy was based almost exclusively on chant melodies: from the earliest organum on Mass Proper chants to the settings of sections of the Mass Ordinary in the fourteenth and well into the fifteenth century, composers built their sacred polyphonic edifices over plainsong foundations.[5] Beginning with Le Rouge and his generation, however, this ancient tradition began to erode as composers chose, with ever-increasing frequency, to compose masses based on vernacular love songs. Why did this shift occur?

Although no history of fifteenth-century music fails to observe the "invention" of the song-based cyclic mass, the genre has been treated primarily as a compositional "type," of interest primarily for its participation in the development of cantus firmus technique and especially its centrality in the development of parody as a compositional device. Thus, despite exhaustive analyses focused on the borrowing techniques of major segments of the repertory, the question of why the chanson mass came into existence in the first place still looms large.[6] Recent studies of the mass that push beyond the description of the overt borrowing of preexistent music to consider the covert presence of the preexistent texts have offered tantalizing glimpses of another dimension of musical borrowing, one in which the dialectic of sacred and profane is played out on a detailed textual level.[7] Still needed, however, is a broadly based investigation of why masses based on vernacular love songs emerged when and where they did, and how they were understood and appreciated in their time.

Certainly the most forthright explanation of the significance of basing a mass on a love song was offered by Howard Mayer Brown: "It may be . . . that song masses and especially those based on courtly songs that celebrate love were normally meant either to be dedicated to the Virgin Mary or to some other beloved saint, following the late medieval inclination to equate secular and sacred love, or else to have been intended originally for weddings, where a celebration of secular love would be most appropriate."[8] Brown's observations, which look through the lens of contemporary social and religious attitudes to bring into focus the practical question of how these masses were intended to function in a liturgical context, provide the starting point for this exploration of why the chanson mass developed when and where it did.

* * *

Clues to the mystery of when and where the chanson mass first emerged appear upon tabulation of the early extant masses, their models, the composers involved, and the sources (see Figure 1.1).[9] Most of the known composers in the earliest layer of this matrix of songs and masses pull us into the Burgundian orbit (with Le Rouge, Binchois, Fontaine, Adrien Basin, and Dufay) and into the circle of the French court (with Barbingant, Faugues, Basiron, and Ockeghem).[10] That an English role is apparent at the inception of the tradition (both Bedyngham and Frye were among the first to compose

Composer	Mass	Song Model/Earliest Source[a]
French-Texted Models		
Bedyngham	*Missa Dueil angoisseux* (à3)	on ballade by Binchois TrentC 88 (ca. 1456) TrentC 90 (ca. 1454–58)
Le Rouge	*Missa Mon cuer pleure*	on rondeau by Fontaine lost
anonymous	*Missa La belle se siet* (à3)	on ballade by Dufay based on popular song TrentC 90 (ca. 1456)
Dufay	*Missa Se la face ay pale* (à4)	on ballade by Dufay TrentC 88 (ca. 1456–62)
anonymous	*Missa Puisque m'amour* (à4)	on rondeau by Dunstable TrentC 88 (ca. 1456–62)
anonymous	*Missa Le serviteur* (à3)	on rondeau by Dufay TrentC 88 (ca. 1456–62)
anonymous (Faugues?)[b]	*Missa Pour l'amour d'une* (à3)	on anonymous lost chanson VatSP B80 (ca. 1458–61)
anonymous	*Missa Esclave puist* (à3)	on rondeau by Binchois TrentC 88 (ca. 1462)
Faugues	*Missa Le serviteur* (à4)	on rondeau by Dufay TrentC 88 (ca. 1462) (att. Ockeghem)
anonymous	*Missa Se tu t'en marias* (à4)	on Binchois "Filles a marier" TrentC 88 (ca. 1462)
anonymous	*Missa Du cuer je souspier* (à3)	on anonymous chanson TrentC 89 (ca. 1460–62)
anonymous (Tinctoris?)[c]	*Missa Nos amis* (à4)	on anonymous chanson TrentC 89 (ca. 1460–62)
Touront	*Missa Monÿel* (à4)	on anonymous lost ballade[d] TrentC 89 (ca. 1462–64)

[a] Sigla of sources from *CC*.

[b] Christopher A. Reynolds, *Papal Patronage and the Music of St. Peter's, 1380–1513* (Berkeley, Los Angeles, and London: University of California Press, 1995), 172–81.

[c] Reinhard Strohm, "Die Missa super 'Nos amis' von Johannes Tinctoris," *Musikforschung* 32 (1979): 34–51, here 39–44.

[d] Adelyn Peck Leverett, "Song Masses in the Trent Codices: The Austrian Connection," *Early Music History* 14 (1995): 205–56, at 238–44, reconstructs the lost model.

Fig. 1.1 Worklist of masses based on love songs, copied before 1477.

anonymous (Busnoys?)[e]	*Missa Quand ce viendra* (à4)	on rondeau by Busnoys TrentC 89 (ca. 1462–64)
Barbingant	*Missa Terriblement* (à3)	on anonymous bergerette VatSP B80 (ca. 1463)
anonymous (A. Basin?)[f]	*Missa Nos amis* (à4)	on chanson by A. Basin LucAS 238 (ca. 1469–70) (incomplete)
anonymous (Basiron?)[g]	*Missa D'ung aultre amer* (à4)	on rondeau by Ockeghem VatSP B80 (ca. 1470)
anonymous (Dufay?)[h]	*Missa Puisque je vis* (à4)	on anonymous chanson VatS 14 (1470s)
Heyns	*Missa Pour quelque paine* (à4)	on anonymous chanson LucAS 238 (1470s) (anon.) BrusBR 5557 (1470s) (att. Ockeghem) VatS 51 (ca. 1474) (att. Heyns)
Caron	*Missa Accueilly m'a la belle* (à4)	on chanson by Caron VatS 51 (ca. 1474)
anonymous (Busnoys?)[i]	*Missa L'ardent desir* (à4)	on anonymous chanson VatS 51 (ca. 1474)
anonymous	*Missa Je ne demande* (à4)	on rondeau by Busnoys VatS 51 (ca. 1474)
Ockeghem	*Missa De plus en plus* (à4)	on rondeau by Binchois VatS 14 (ca. 1470s)
anonymous (Martini?)[j]	*Missa Au chant de alouette* (à4)	on popular song VatSP B80 (ca. 1473)
anonymous	*Missa Le serviteur* (à3)	on rondeau by Dufay TrentC 89 (ca. 1462–64)

[e] Reinhard Strohm, *Music in Late Medieval Bruges* (Oxford: Clarendon, 1985), 128.

[f] Ibid.

[g] Lawrence Feininger suggested Basiron's authorship (Josephus [José] M. Llorens, *Capellae Sixtinae Codices musicis notis instructi sive manu scripti sive praelo excussi*, Studi e testi 102 [Vatican City: Biblioteca Apostolica Vaticana, 1960], 20). Reservations about this attribution and the date of the mass are raised by Rob C. Wegman, "The Anonymous Mass *D'Ung aultre amer*: A Late Fifteenth-Century Experiment," *Musical Quarterly* 74 (1990): 566–94, here 591–93.

[h] Laurence Feininger, ed., *Missae de G. Dufay et Anonymi*, Monumenta Polyphoniae Liturgicae Sanctae Ecclesiae Romanae I/2 (Rome: Societas Universalis Sanctae Ceciliae, 1952), Mass 4.

[i] Attribution first proposed in Llorens, *Capellae Sixtinae Codices*, 104, on Feininger's authority; subsequently endorsed by Rob C. Wegman in "Another Mass by Busnoys?" *Music and Letters* 71 (1990): 1–19.

[j] Reynolds, *Papal Patronage*, 238–46.

Fig. 1.1 (*cont.*).

English-Texted Models		
Le Rouge	*Missa Soyez aprantiz* (à3)	on ballade by Frye ("So ys emprentid") TrentC 90 (ca. 1454–56)
Frye	*Missa So ys emprentid* (à3)	on ballade by Frye LucAS 238 (ca. 1467–69?) (incomplete)
Italian-Texted Models		
anonymous (Joye?)[k]	*Missa O rosa bella II* (à3)	on ballata by Dunstable or Bedyngham TrentC 90 (ca. 1456)
anonymous (Joye?)[l]	*Missa O rosa bella I* (à3)	on ballata by Dunstable or Bedyngham TrentC 88 (ca. 1462)
anonymous	*Missa O rosa bella III* (à4)	on ballata by Dunstable or Bedyngham TrentC 89 (ca. 1462–64)
anonymous	*Missa Gentil madona mia* (à4)	on song by Bedyngham[m] TrentC 91 (ca. 1470)
German-Texted Models		
anonymous	*Missa Groß Sehnen* (à4)	TrentC 89 (ca. 1462)
anonymous	*Missa Deutscher Lieder* (à3)	TrentC 89 (ca. 1462–66)
anonymous	*Missa Wünchlichen schön* (à3)	TrentC 89 (ca. 1465–70)
anonymous (Touront?)[n]	*Missa Sig säld und heil* (à4)	TrentC 89 (ca. 1462–64) (incomplete) TrentC 91 (ca. 1465–70)
anonymous	*Missa Grüne Linden* (à3)	TrentC 88 (ca. 1462)
Philipus	*Missa Hilf und gib rat* (à4)	TrentC 89 (ca. 1464)
Vincenet	*Missa Zersundert ist das junge Herze mein* (à3)	TrentC 91 (ca. 1465)
anonymous	*Missa Zersundert ist das junge Herze mein* (à4)	TrentC 91 (ca. 1468)

[k] Strohm, *Music in Late Medieval Bruges*, 29.

[l] Ibid.

[m] David Fallows, *A Catalogue of Polyphonic Songs, 1415–1480* (Oxford: Oxford University Press, 1999), 500.

[n] Leverett, "Song Masses," 244–47.

Fig. 1.1 (*cont.*).

masses based on songs, and Frye's ballade *So ys emprentid* served as an early model) is not surprising, given the frequency of Burgundian and English contact during the reign of Philip the Good (1419–67).[11] So although we will never know the precise provenance of these works, it seems reasonable to start from the premise that northern court culture provided the cradle for the production and consumption of song-based masses at the beginning of their cultivation.[12] This is hardly surprising: who could better apprehend and appreciate the crafty inclusion of their favorite love songs within music for the mass than the ladies and gentlemen of the courts for whom the songs themselves were created?

What might surprise is, first, the rapid diffusion of the chanson mass, whose earliest extant sources derive from Trent, Naples, and Rome, and second, the immediate appeal the genre apparently held for clerical as well as lay worshipers, and ecclesiastical as well as courtly institutions. The manuscripts from Trent that transmit the earliest layer of this repertory, for example, were copied in large part by the clergyman Johannes Wiser, who served first as rector of the cathedral school in Trent and then as chaplain to the Bishop of Trent;[13] the choirbook VatSP B80 was made for use by the choir of St. Peter's Basilica in Rome.[14] So while we may seek an enhanced understanding of the chanson mass within the context of the northern court culture in which it blossomed, the joining of secular and sacred elements in this most serious of liturgical genres was implicitly accepted if not condoned in less worldly environments.

While the sources reveal the widespread appeal of the chanson mass, they tell little about the intended function of these works within the liturgy. The nuptial context suggested by Brown rests solely on the generally accepted hypothesis that Dufay's *Missa Se la face ay pale*, based on a chanson believed to date from the composer's years at the court of Savoy during the 1430s, was destined for a Savoyard marriage celebration, either that of Carlotta of Savoy to the future Louis XI of France in 1451 or the consummation of the marriage of Amadeus of Savoy and Yolande de France in 1452.[15]

More promising, in terms of furnishing a springboard from which to launch a consideration of the cultural framework within which the chanson mass grew, is Brown's reference to "the late medieval inclination to equate secular and sacred love." The sources do support the suggestion that songs originally directed to a courtly lady were, if cloaked in liturgical dress, appropriately redirected to the Queen of the heavenly court, Mary. Provision of complementary motets on Marian texts over the same song tenors that underlie several of the early chanson masses (the anonymi *Missa Soyez aprantiz* and *Missa Esclave puist-il devenir*) point to a Marian context for these works.[16] Further Marian associations are evident within the slightly later German song-based mass tradition: in TrentC 91 a polyphonic Marian introit, *Salve sancta parens,* immediately precedes the *Missa Sig säld und heil,* which itself incorporates Marian plainsongs for the Kyrie and Sanctus in those movements.[17]

This penchant for juxtaposing secular with sacred love, courtly lady with Blessed Virgin, is a constant throughout the medieval period; it is manifest in literary, visual, and musical creations for more than three centuries before the rise of the chanson mass. A brief exploration of the origins and development of this correspondence across a range of expressive media prior to the fifteenth century will paint a cultural landscape within which we can begin to situate masses based on secular love songs.

* * *

During the three centuries prior to the rise of the chanson mass, fascinating and complex interpretive traditions developed in theology and vernacular literature around the topics of sacred and profane love, the Virgin and the Lady. The main catalyst behind these juxtapositions is clearly the Old Testament Song of Songs, with its ambiguous male and female voices expressing their mutual love in passionately erotic terms. Christian theologians were drawn to the poem early on: in the third century Origen wrote two commentaries that established the tradition of interpreting the Song as an epithalamium, a nuptial song between Christ the bridegroom and the soul or Church as bride. Jerome, Gregory the Great, and Bede are just three of many early medieval exegetes to follow Origen in attempting to understand the Song in these terms.[18]

Direct association of the beloved in the Song of Songs with the Blessed Virgin was first made not within the commentaries, but within the liturgy itself.[19] As early as the seventh century, liturgies for the feasts of the Nativity, the Purification, and especially the Assumption of the Virgin harnessed Solomonic imagery and language in praise of the Virgin.[20] Throughout the Middle Ages, Marian liturgies across Europe were increasingly peppered with plainsongs based on lines from the Song: "Sicut lilium inter spinas," "Hortus conclusus," "Quam pulchra es amica mea," "Nigra sum sed formosa."

Only in the twelfth century did theologians undertake commentaries developing the Marian exegesis of the Song. The enormous interest in the poem that burgeoned then speaks to a culture's increasing desire for a loving, maternal agent of intercession most acutely expressed in monastic circles. Twelfth- and thirteenth-century scholastic commentators such as Honorius Augustodunensis, Hugh of St. Victor, Rupert of Deutz, Alan of Lille, and Bernard of Clairvaux developed interpretations of the Song as a portrait of the love between God and the Virgin Mary, who becomes simultaneously mother and spouse of Christ, personified as a beautiful young woman pleading in ardent language for the kiss of her beloved son and husband.[21] Vernacular commentaries on the Song early on betray the commingling of the worlds of cloister and court; a twelfth-century version from northern France, for example, is presented as a *roman,* dedicated to a lady, with interlocutors addressing one another as "Damoiselle," "amie," and "Bels sire."[22]

At the same time, the erotic imagery of the Song imbued vernacular love poetry, furnishing stock literary phrases for expressions of adoration, desire, and beauty.[23] Thus twelfth-century troubadours and trouvères expressed the ideals of courtly love in verse whose motifs and attitudes are duplicated in Latin religious lyric poetry; courtly lady and Blessed Virgin alike are described as beautiful, passive, and noble examples of femininity to be won by ideal, chaste "love from afar."[24] This allegorical interpretation of the Song of Songs, which at once humanizes the Blessed Virgin and elevates carnal woman, resonates in vernacular literature of the thirteenth century, both in the romance genre, such as the *Roman de la Rose,* and in loftier allegories such as Dante's *Divine Comedy.*[25] A small treatise on the nature of love by the mid-thirteenth-century Cistercian monk Gérard de Liège exemplifies the blurred boundaries between the domains of sacred and secular love at this time: Gérard composes a pious and orthodox encomium to divine love in Latin prose in which the main points are made by quoting the French lyrics of secular love songs, as this excerpt shows:[26]

Ipsa anima mansuetior ad correctionem, inde patientior ad adversitatem et laborem, inde sagatior ad cautelam, inde ardentior ad amorem, inde humilior pro conscientia, inde acceptior et magis placens pro verecundia, inde paratior ad obediendum, inde ad gratiarum actionem devotior ac sollicitior. Et hec dicit Bernardus.	The soul becomes more yielding to correction, and thus more patient in adversities and trials, more prudent in its caution, more ardent in love, humbler by virtue of its sensibility, more acceptable and pleasing by virtue of its modesty, readier to obey and more devoted and solicitous in giving thanks. And this is what Bernard says.
Unde talis anima bene potest cantare quoddam carmen quod vulgo dicitur:	Such a soul, therefore, can indeed sing a song which is heard everywhere:
Grevet mout li mal damours, mius en vaurai, Car plus sages en serai, Et de foliser allours, me garderai.	*The malady of love causes great grief, yet I shall be the better for it, for I shall be wiser through it, and guard myself against dallying elsewhere.*

Visual representations of the Virgin during the twelfth and thirteenth centuries respond as well to the sensualization and humanization of Mary inspired by the Song, introducing themes and attitudes expressing her embodiment of the female virtues of humility, submissiveness, and tender motherhood.[27] A comparison of French cathedral sculpture by Penny Schine Gold is revealing.[28] Above the west portal at Senlis Cathedral, carved ca. 1170, Mary sits crowned and upright beside Christ the King, her posture regal and aloof. But at Chartres, some twenty-five years later, the same iconographic theme of the Coronation of the Virgin adorning the north portal tympanum renders her with head submissively inclined and hands gesturing in supplication.

Similarly, the west portal at Amiens Cathedral, carved ca. 1220–1230, presents a standing Virgin and Child, the mother impassive and disengaged from an undemonstrative baby, both gazing directly at the viewer. At the south door of the same edifice, the identical theme chiseled some twenty-five years later portrays a tender young mother tilting her smiling face toward the baby perched on her swaying hips, each gazing lovingly at the other. The proliferation of iconographic themes that focus on Mary's sensual maternity (the nursing mother, *Maria lactans*) and on her emotional capacity (the grieving mother, *Mater dolorosa*) further demonstrate the increasing emphasis on her humanity.[29]

Composers too availed themselves of the expanded vocabulary of adoration for both courtly lady and Blessed Virgin. The pious chansons of the troubadours and trouvères proclaim a joyous love rapture different from secular love songs only in the object of their affection; language and musical style are indistinguishable from their secular counterparts.[30] Sequences are directed in ever-increasing number to the Blessed Virgin and indulge a sensual language indebted to the Song of Songs.[31] Perhaps most self-conscious are the juxtapositions of sacred and profane love that appear in the genre of the thirteenth-century polyphonic motet, juxtapositions that invite richly textured allegorical interpretations through the combination of tenors based on snippets of Marian chant, erotic vernacular lyrics, and devotional Latin texts.[32] And as the apposition of courtly lady with Blessed Virgin and sacred with carnal love continues in Latin and vernacular literature and the visual arts during the fourteenth century, so do its musical manifestions: Guillaume Machaut, for example, exploited these tensions with great subtlety in several of his motets.[33]

* * *

The fifteenth century thus inherited a well-established interpretive tradition based on constructing analogies between carnal and spiritual love, courtly lady and Mother of God. It was a tradition developed and refined by scholastic theologians for whom analogy was essential because it allowed the human mind to rationally comprehend the divine through imperfect comparisons. In other words, imperfect secular love was understood to participate, in the Platonic sense, in perfect spiritual love; the Lady resembled, albeit imperfectly, the Blessed Virgin.[34]

This analogical framework is obviously germane to a general understanding of the juxtapositions and fusions contained in much medieval art, literature, and music. One senses, however, a qualitative difference in the fluid transposition of a biblical language of love to the secular arena, the general humanization of Marian imagery, and the playful ambiguity of nonliturgical motets combining materials concerned with sacred and profane love as compared with the bold act of basing the most ambitious musical adornment for the most solemn ritual of the Catholic Church on unambiguously secular love songs. Both the scale and the technique—effectively claiming an entire secular genre for use as a liturgical trope[35]—set the chanson mass apart from earlier combinations or juxtapositions of secular and sacred love in musical, literary, or visual media. And indeed, new trends in fifteenth-century northern culture, of which the chanson mass is one manifestation, suggests that there are other influences and concerns at work.

These new tendencies are captured most dramatically in the devotional images of the Blessed Virgin created by such artists as Jan van Eyck, Rogier van der Weyden, and Hans Memling for aristocratic and well-to-do bourgeois patrons. As never before, viewers are invited into a direct, intimate, and emotional relationship with the Mother of God, whom they are invited to imagine as a contemporary courtly lady, idealized paragon of beauty and virtue on earth. She becomes, during the first half of the fifteenth century, not only approachably human but invitingly familiar. Compare, for example, two treatments of the Annunciation scene. An illumination from a Book of Hours made ca. 1415 by a follower of the Boucicaut Master sets the event in its traditional venue, the interior of a church; Mary's covered, haloed form retains a certain distance (Figure 1.2). Two decades later, a panel painting by Rogier van der Weyden presents the viewer with a Virgin who has literally let her hair down, abandoned the halo, and stepped into a bedroom in a Flemish townhouse; remove the angel, and the scene might be construed as an intimate glimpse into a well-bred Flemish lady's private devotions (Figure 1.3).[36]

The *Maria lactans* theme is similarly transformed. At the end of the fourteenth century, an illumination for *Les Très Belles Heures* made by Jacquemart de Hesdin for the Duke of Berry portrays a gently smiling Madonna who is nonetheless inward-looking; a halo encircles her head, and she seems unaware of the Child at her breast. The donor portrait of the duke on the

Fig. 1.2 "Annunciation" from a Book of Hours made ca. 1415 by a follower of the Boucicaut Master (New York, The Pierpont Morgan Library, MS M.455, fol. 23); used with permission.

Fig. 1.3 "Annunciation," panel painting made ca. 1435 by Rogier van der Weyden (Paris, Musée du Louvre); used with permission.

facing page, although looking at the image of the Virgin, clearly occupies a separate space (Figure 1.4). Some forty years later, an altarpiece by Rogier van der Weyden portrays her as a fond young mother nursing her happy infant, part of a Flemish townscape that includes not only St. Luke but two small and very human figures enjoying the view in the center background (Figure 1.5).

The representation of the Blessed Virgin as a contemporary lady persisted in northern art throughout the fifteenth century; the ease with which she joined the courtly environment is perhaps best demonstrated in paintings depicting the Virgin Among Virgins. The Master of the St. Lucy Legend, for example, painted the "Virgin and Child with Female Saints in a Rose Garden" sometime before 1483; the attire of these virgin saints is the height of court fashion, and the towers of Bruges loom in the background. These

Fig. 1.4 "The Duke of Berry presented to the Virgin," from *Les Très Belles Heures,* made ca. 1400 by Jacquemart de Hesdin (Brussels, Bibliothèque Royale, MS 11060–61, fols. 10v–11); used with permission.

saintly ladies are shown in a private garden enjoying the domestic pleasures of well-bred women: reading, sewing, playing with a child (Figure 1.6). And it is but one small step from imagining the Blessed Virgin as the generic courtly lady to representing the Virgin as a particular noble and beloved woman; thus the right panel of Jean Fouquet's Melun Diptych (ca. 1450) presents a Virgin and Child in which the Virgin, impeccably and fashionably garbed and groomed, is believed to be a commemorative portrait of Agnès Sorel, mistress of King Charles VII, who died in 1450 (Figure 1.7).[37]

How are we to understand this new enthusiasm for representing the Virgin Mary in such mundane contexts? Is this, as the cultural historian Johan Huizinga would have it, symptomatic of late medieval society's moral decay, signaled by its "fatuous familiarity with the sacred" and "dangerous association of religious with amatory sentiments"? Is the appearance of the chanson mass, with its shameless use of a secular genre as a liturgical trope, merely the flip side of the same coin, both revealing a decadent compulsion to uncontrolled analogy that all but obliterates the demarcation between religious and worldly and inevitably invites irreverence?[38] The analogical

Fig. 1.5 "St. Luke Painting the Virgin," panel painting made ca. 1440 by Rogier van der Weyden (Boston, Museum of Fine Arts); used with permission.

model as a useful means for understanding the divine through the mundane does seem in these examples to have reached the point of collapse.

Indeed, the conceptual environment from which the period's religious art, devotional literature, and sacred music emerged is heavily inflected by intellectual and devotional trends that both enrich and erode the power of

Fig. 1.6 "Virgin and Child with Female Saints in a Rose Garden," panel painting made before 1483 by the Master of the St. Lucy Legend (Detroit, The Detroit Institute of Arts); used with permission.

Fig. 1.7 *Melun Diptych,* right panel, made ca. 1450 by Jean Fouquet (Antwerp, Koninklijke Museum voor Schone Kunsten, Inv. 132); used with permission.

analogy to explain such conflations of sacred and secular. These trends are exemplified by the mystical theology of Jean Gerson (1363–1429), the foremost northern theologian and church leader at the opening of the fifteenth century, chancellor of the University of Paris and Burgundian familiar.[39]

Gerson, through voluminous writings directed to a broad lay as well as clerical audience, rejected the scholastic claim that humans could rationally comprehend the divine. Instead, he encouraged affective contemplation as the means to experience the rapture of mystical love, through which humans could reach God. For Gerson, God was unknowable through reason, and thus analogy becomes a useless tool for the soul in search of God. The experience of mystical love, Gerson argued, is beyond rational understanding and expression.[40]

So in order to communicate the essence of the sublime experience of mystical love, Gerson appeals to the affective realms of human experience using language appropriate to those realms. Thus, in *La Montaigne de contemplation,* written in 1400 and directed to his sisters and other women, Gerson describes this process of being drawn to God through love in terms of ordinary worldly love: the person caught in *amour mondaine* languishes from desire for what she loves madly, becoming so devoted to the desired object that all else ceases to exist. Gerson then presents this obsessive love as a paradigm, a kind of model or template, for spiritual love, in which the lover forgets the world in her all-consuming love of God. Thus it is the movement to a state of rapturous transport, in which all but the beloved is forgotten, that is the same in both carnal and spiritual love; the nature of the object of desire is, however, completely different.[41]

Moreover, Gerson often invokes the extravagant poetic conventions of courtly love to capture the nature of spiritual love, as in *Le Jardin amoureux de l'ame.* The passage below reveals him playing with the courtly "garden of love" *topos,* itself rooted in the Song of Songs, from which he quotes. Christ is personified as the loyal (and musically inclined) lover beckoning his beloved:

> En ce mondain desert est le jardin d'amoureuse consolacion ou le vray Dieu d'amours habite; c'est le jardin gracieux ou habite le doulx Jesus et ouquel il appelle sa mie quant il dit ou livre de chansonnettes amouresuses: Veni in ortum meum, soror mea, sponsa mea. Viens, dit il, en mon jardin, ma doulce suer et ma chiere espouse. Ceste doulce voix et ceste chansonnette amoureuse chante melodieusement Jhesucrist, le loyal amant, en appellant a soy la sainte ame qu'est de luy amouree par ardent charité; et la nome sa suer et son espouse; sa suer par consanguinité de nature, et son espouse par affinité de grace; sa suer par la semblance de nature humaine que il prist en la vierge Marie, et son espouse par la beauté de grace divine qu'elle prent de Dieu le pere; sa suer par le lignage naturel, et son espouse par mariage espirituel. Helas, bien doit celle estre louee et benoicte qui est de si grant lignage et en si hault mariage qu'elle est nomé suer et espouse du grant roy du ciel et du haut empereur de monde. Vecy doncques petites parolles et pleines de grant sentence, embrasee de ardent amour et arrousee de amoureuse doulceur.[42]

The "chansonette amoureuse" that concludes this tract, sung by the enraptured soul to Jesus Christ, adopts wholesale the extravagant poetic style of the *grands rhétoriqueurs:*[43]

> Pour la couronne amoureuse gagner
> Dont aux amants fait amoureux present
> le Dieu d'amour, tous le doivent louer,

Amer, servir tres amoureusement
Car par l'amour de sa volonte pure
par la doulceur de sa franche nature
Fu toute riens en son art pourpensee
Et de nyent a sa forme formee
Toute creature.[44]

The chanson mass can be heard as a musical counterpart of Gerson's appeal to affective experience and courtly language within the context of his religious writings, the sonorous conjuring of the secular song within the mass effecting a similar meshing of expressive elements usually confined to separate spheres. In the chanson mass as in Gerson's theology, a codified poetical/musical language of secular love suggests to the reader/listener the paradigmatic relationship of his or her personal experience of corporeal love to the mystical experience of divine love. In other words, worldly love is invoked, through the use of song material, to suggest the transport of carnal love; that this carnal world is forgotten through the transport of spiritual love is suggested by the fact that the secular model is largely forgotten, as an audible entity, within the liturgical setting.

Thus, in the chanson mass as in the devotional tracts of Gerson, the expressive language of the courtly love lyric serves naturally and appropriately, indeed inevitably, to address a divine beloved. Religious poetry composed in the same courtly environments that fostered the chanson mass likewise recognizes the dual utility of the poetic language. A poem by Jean Molinet, official chronicler at the court of Burgundy from 1475 until 1507, and active in that sphere from at least the 1460s, vividly illustrates this duality.[45]

Molinet's poem carries the rubric "A poem that may be addressed either to the Virgin Mary or by a lover to his lady" ("Dictier qui se poeult adreschier soit a la vierge Marie ou pour un amant a sa dame").[46] Here Molinet invokes the praise register of the courtly lyric; in the opening verse, beauty, nobility, and various virtues characterize both courtly damsel and Blessed Virgin Mary; the attribute of loyalty is stressed by the refrain:[47]

Dame sans per, tres precieuse image,
Chief d'oeuvre exquis et des dieux l'artifice,
Fleur de beaulté, de noble et hault lignage,
A qui chascun doit tribut et homaige,
Salut, honneur, loenge et sacrifice,
Il n'est pappier, ne plume qui souffise
A descripre vostre bonté totalle,
Car vous estes en coeur tousjours lealle.

The eighth verse focuses on an enumeration of the beloved's physical attributes—head, body, eyes, mouth, and brow:

De tous biens plaine et de beaulté l'exemple
Estes sus terre et nul ne vous en passe;
Vous exedés toute imaige de temple
De chief, de corps, d'yeux, de bouche et de temple;
La plus des plus estes et l'outrepasse;
Toute beaulté que nature compasse
N'est emprés vous sinon obscure et palle,
Et vous estes clere et tousjours lealle.

Molinet's attention to the details of physical appearance, natural indeed to erotic love, proves characteristic of Marian devotion at this time as well: this very preoccupation with the details of the appearance of the Virgin underlies a curious Marian devotion known as the "Benediction of the Virgin's Limbs" that flourished in the north during the fifteenth century. Devotional manuals prescribed various benedictions and devotions directed to individual parts of the Virgin's body, including lips, mouth, ears, teeth, nose, neck, and breasts, to be performed before her image.[48] At the root of this devotion and Molinet's poem lies the Mariological tradition of Solomonic commentary, but the emphasis on private devotional contemplation effected by reading such poems and manuals serves most immediately to encourage the rapture of mystical love through the paradigm of erotic love's rapturous transport.

The eighth verse of Molinet's poem begins with the phrase "De tous biens plaine" ("Full of all good"), clearly a reference to the poem set to music by Hayne van Ghizeghem, a composer in the service of Charles the Bold.[49] With this quotation the poet's reliance on the language of courtly love is made explicit, just as the composer's reliance on the musical embodiment of that language is made explicit through the musical quotation of a song within a mass. Both kinds of quotations obviously presuppose an audience well acquainted with the chanson repertory, and so serve to trigger an associative reaction in reader or listener. In both cases the quotations conjure generic expressive registers—praise, plea, lament—easily transposed to the sacred domain.

Molinet used this device of chanson quotation most extensively in a poem directed explicitly to the Blessed Virgin. The *Oroison a Nostre Dame* of ca. 1467 is fashioned of nineteen verses of eleven lines each; the first and last line of each verse quotes the incipit of a chanson, for a total of thirty-six song citations. The poem begins as follows:[50]

Tout a par moy, affin qu'on ne me voye,
Me suis retraict en ce lieu solitaire,
Considerant que vous estes la voye

Ferme et constante ou tout coeur se ravoye
Et de vertu le lieu proprietaire;
Pourquoy, Marie, a qui je doibs complaire,
Vostre plaisir soit tel que je vous serve,
A celle fin qu'en servant je desserve
Ce qu'ung leal amant doit deservir,
Miroir divin, melliflue conserve,
Vecy l'amant qui vient pour vous servir.

Je ne vis oncques la pareille
De vous, gratieuse pucelle;
Pourtant clinés vers moys l'oreille,
Car a vous servir m'apareille,
Comme de toutes la plus belle,
Mere de Dieu, vierge et ancelle,
Gardin fermé, clef davidicque,
Sourse d'honneur, trosne ivoricque
Assis par dessus les noeuf cieux;
Navré suis mieux que d'une picque
Par le regard de vos beaux yeux.

The formal strategy here correlates with that of most chanson masses: from the earliest examples of the genre, composers usually conjure the song most forthrightly at the beginning and/or end of major sections through the use of parody (most familiar would be Dufay's *Missa Se la face ay pale*, in which the memorable final melisma appears at the end of every movement). Parody as a contrapuntal technique involves the musical ingestion of a snippet of the original polyphony into a new context; in the chanson mass, as in this poem, such ingestion of song segments might also be regarded as a process of musical transmogrification, turning mundane into divine.

Six of the chansons from which Molinet borrowed to fashion this courtly salute to the Virgin also appear garbed in sacred dress within the early chanson mass. All articulate a distinct expressive register, striking a certain poetic/musical posture toward the beloved, whether damsel or Blessed Virgin Mary: the lament register (*Se la face ay pale, Dueil angoisseux, Pour quelque paine*), the plea register (*Le serviteur, D'ung aultre amer*), and the subservient lover (*Esclave puist il devenir*).

Earthier declarations of love were also redirected to the divine paramour, both in Molinet's poem and in the chanson mass. Molinet begins the tenth verse of his *Oroison* with the opening line of the song *Allegiez moy, doulce plaisant brunette* (Soothe me, sweet pleasing brunette); the poet continues "Ou Jesus Crist volt prendre char humaine" (through which Jesus Christ assumes human form), but the sense of the original chanson brought to the reader's mind by this quotation was unabashedly bawdy (Soothe me, sweet pleasing brunette, *desous la boudinette*, just below the belly button). Likewise, masses occasionally are based upon chansons that cast the lady as fickle coquette (*Missa Puisque m'amour*) or as cruel and heartless (*Missa Accueilly m'a la belle*). But this need not disturb us. A powerful aesthetic of play and delight in concealment is integral to the courtly culture in which the chanson mass developed; the elite audience for Molinet's poetry and the chanson masses clearly prized such in-jokes.[51] Both Molinet and the composers are having fun, but all assume that their readers/performers/listeners know the secular original and have enjoyed its humor, even its titillation. The intention is clearly not to shock or to be irreverent—the readers/performers/listeners are invited both to be amused and to contemplate the transformation undergone by the secular original in its new context. Molinet himself, explaining his motivation for moralizing the *Roman de la Rose*, wrote that he endeavored to "convert with my rough millstone the depraved into the virtuous, the corporeal into the spiritual, the worldly into the divine").[52]

* * *

Conjuring and transforming the sounds and expressive registers of the chanson within the mass thus appears to serve a semiological function akin to the use of courtly language in Gerson's religious writings and Molinet's religious poetry, akin to the depiction of the Virgin in courtly dress or familiar environment. The contemplative courtly worshiper's goal, facilitated through the presence of such potent musical, linguistic, and visual signs, is vividly suggested in an extraordinary miniature in the Hours of Mary of Burgundy, from ca. 1477 (Figure 1.8). Mary of Burgundy herself, garbed in the height of court fashion, is shown reading at an open casement window; through this contemplative activity she is transported in her imagination, even more richly robed and with several ladies in waiting, into the presence of the Virgin Mother with infant Christ, where the Virgin Mary gazes directly into her upturned eyes. The secular world that frames this scene, perhaps symbolized by the lavish pendant on the windowsill, laid aside by the intent reader, falls away as the viewer is drawn to the central image of the Blessed Virgin. While a prayer from a Book of Hours here provides the mystical catalyst that so transports her, Gerson's religious tracts, Molinet's religious poetry, or even a devotional picture would serve as well, and it is not hard to imagine the church choir filled with the sounds of a chanson mass.

In sum, masses based on love songs, like the devotional pictures, religious tracts, and writings presented here, can be usefully understood as the courtly expression of new trends in religious thought and Marian worship that go beyond and in some crucial ways contradict the analogical model that dominated ways of thinking prior to the fifteenth century. Binding these diverse modes of expression together is a common reliance on the language

Fig. 1.8 "Mary in Church," from The Hours of Mary of Burgundy, made ca. 1477 by the Master of Mary of Burgundy (Vienna, Österreichische Nationalbibliothek, Cod. 1857. fol. 14v); used with permission.

of secular love, be it musical, verbal, or visual, as the communicative vehicle best suited to transport the listener, reader, or viewer into contact with the divine.

* * *

That courtly environments continued to provide the most comfortable climate for the cultivation of chanson masses well into the sixteenth century is suggested by the mass repertories of composers such as Josquin, Obrecht, Crequillon, and Gombert, many if not all of whose song-based masses are likely to have been initially intended for performance by court chapels. But the practice touched nerves then and now. Although song-based masses continued to be composed and widely distributed well into the sixteenth century, objections to the use of secular materials in the mass did surface. Conrad von Zabern may have voiced the earliest opposition to the use of songs in masses; in his *De modo bene cantandi* of 1474 he rails against the use of secular songs in the Gloria, Credo, Sanctus, and Agnus Dei, claiming they scandalize the faithful and evoke the dance hall.[53] In 1517 the authorities at Cambrai Cathedral ordered the removal of the *Missa Comment peult avoir joye* from their music books, possibly reflecting growing concern about the use of secular songs in church expressed in ecclesiastical councils of the period.[54] The demise of the song mass tradition coincides, not surprisingly, with the period of the Council of Trent, whose legislation on liturgical matters effectively proscribed such practices.[55]

The historical study of the polyphonic mass since the mid-nineteenth century reveals comparable reservations about the inclusion of secular love songs within liturgical polyphony. Huizinga's suspicion of the chanson mass, expressed in condemnatory terms reminiscent of the various ecclesiastical councils, has already been noted. And although his forthrightly moralistic attitude does not appear in subsequent treatments of the genre by historians of music, these seem to admit their hesitations about the chanson mass by simply avoiding the question as to why composers chose to incorporate chansons into their masses. Indeed, Wilhelm Ambros's mid-nineteenth-century formulation of the cyclic mass as a genre that transcended its function and materials through the composers' creative genius made the question irrelevant. Ambros characterized the song material within masses exclusively as thematic and contrapuntal fuel, the "formula at the foundation" of the mass, whose materials are, compositionally speaking, "everywhere and nowhere."[56] Even Peter Wagner, whose magisterial study of the mass is exceptionally well attuned to questions concerning the liturgical and social context of the genre, felt compelled to argue that secular cantus firmi were "neutral characters" divested of their content through the omission of their original text.[57] And while the more recent application of the rhetorical concept of *imitatio* to music, which emphasizes the didactic, competitive, and emulative value attached to reworking existing compositions of the period, may partially explain the technical attraction of the chanson mass as an arena of compositional endeavor, it does not address the question of just what masses based on love songs signified to those who made, sang, and heard them.[58]

Only by raising our eyes from the notes on the page and looking as closely at the period's theology, literature, and visual creations can we begin to engage these broad questions of meaning. Having then gained some sense of the larger cultural context within which the chanson mass developed, it is of course essential to return to the notes; analysis too should be informed by consideration of why composers chose certain models. It may prove interesting, for example, to systematically examine how composers treat that single textual phrase of the Mass Ordinary that directly mentions the Virgin Mary ("Et incarnatus est de Spiritu Sancto ex Maria Virgine" from the Credo) or to consider the "inaudible polytextuality" of the song text in relation to the words of the Ordinary.[59] Clearly, composers borrowed preexistent music for reasons that reach beyond the pitches and rhythms they plundered, and to attempt to fathom their reasons can only enrich our understanding and experience of their "new" creations.

Notes

1. Versions of this paper were presented at the 29th International Congress on Medieval Studies at Kalamazoo, Michigan, May 1994; at the University of Wisconsin, Madison, February 1997; and at the 16th International Congress of the International Musicological Society in London, August 1997. I acknowledge with gratitude the help and suggestions of a great many colleagues in musicology as well as art history and literary studies over the course of this project; particular thanks go to Jessie Ann Owens, Bonnie Blackburn, Margaret Bent, Gregory Bloch, Stephen Tifft, Karen Swann, and Laura Weigert.
2. On Fontaine, see Craig Wright, "Fontaine, Pierre," *New Grove II*. The poem begins:

   Mon cuer pleure, mès des yeulx me fault rire;
   C'est pour celler mon doloreux martire,
   Car je n'ose raconter ma clamour.
   Ne nul samblant faire de ma dolour
   A la belle pour qui souvent souspire.

   The rondeau is edited by Nigel E. Wilkins in *A 15th-Century Repertory from the Codex Reina (Paris, Bibl. nat., nouv. acq. Fr. 6771)*, CMM 37 ([Rome]: AIM, 1966), 34–35.
3. A Guillaume Rubi is documented in the Burgundian chapel from 1415 to 1419 and from 1430 to 1450; a Guillaume Le Rouge served in the chapel of Charles d'Orléans from 1451 to 1465. For the most recent summary of the biographical details of persons named Le Rouge, see David Fallows, *A Catalogue of Polyphonic Songs, 1415–1480* (Oxford: Oxford University Press, 1999), 718.
4. Johannes Tinctoris, *Opera theoretica*, ed. Albert Seay, Corpus scriptorum de musica 22 (N.p.: AIM, 1975–), 2a:47.
5. Prior to the emergence of the chanson mass, only two isolated instances of masses incorporating secular material survive. The anonymous fourteenth-century Mass of Tournai employs an amorous French text (not the music) within the Ite missa est; recent research suggests that this mass was intended for votive Marian masses (see Chanoine Jean Dumoulin et al., *La Messe de Tournai: Une messe polyphonique en l'honneur de Notre-Dame à la Cathédrale*

de Tournai au XIVe siècle (Tournai: Archives du Chapitre Cathédral, and Louvain-la-Neuve: Université Catholique de Louvain, 1988). The only composer to include the music of love songs within masses before the rise of the chanson mass was Antonius dictus Zacharias de Teramo (fl. ca. 1400), who based two Gloria settings on his own ballatas *Rosetta che non cambi mai colore* and *Un fior gentil m'apparse* (both apparently Marian). See Kurt von Fischer, "Bemerkungen zur Überlieferung und zum Stil der geistlichen Werke des Antonius dictus Zacharias de Teramo," *Musica Disciplina* 41 (1987): 161–82, esp. 178–79.

6. The seminal investigation of borrowing techniques in fifteenth-century masses is Edgar H. Sparks, *Cantus Firmus in Mass and Motet 1420–1520* (Berkeley and Los Angeles: University of California Press, 1963; reprint ed., New York: Da Capo, 1975), esp. 136–90 passim. Subsequent studies treating borrowing procedures in chanson masses include J. Peter Burkholder, "Johannes Martini and the Imitation Mass of the Late Fifteenth Century," *JAMS* 38 (1985): 470–523; Murray Steib, "Imitation and Elaboration: The Use of Borrowed Material in Masses from the Late Fifteenth Century," (Ph.D. dissertation, University of Chicago, 1992); Cathy Ann Elias, "Imitation, Fragmentation, and Assimilation of Chansons in the Masses of Gombert, Clemens, and Crecquillon: A Kaleidoscopic Process," (Ph.D. dissertation, University of Chicago, 1994); Fabrice Fitch, *Johannes Ockeghem: Masses and Models*, Collection Ricercar 2 (Paris: Honoré Champion, 1997).

7. Michael Long broke important new ground in his study of Josquin's *Missa di dadi* (*Missa N'aray je jamais mieulx*) by revealing that the composer invested the amorous text of the model (not heard within the mass) with a new, religious connotation that determines the manipulation of the model's musical content within the mass (Michael Long, "Symbol and Ritual in Josquin's *Missa Di Dadi*," *JAMS* 42 [1989]: 1–22). Christopher Reynolds likewise recognized that the citation of or allusion to song melodies within masses implicitly involved the songs' texts, attributing the use of chanson tunes in the mass to the rhetorical impulse, sparked by Italian humanists, "to discuss divine matters in the popular language," so as to both "please as well as teach" through "the complex interactions that were possible between the texts (whether written or remembered)" (Christopher A. Reynolds, "The Counterpoint of Allusion in Fifteenth-Century Masses," *JAMS* 45 [1992]: 228–60, esp. 248–52). While I will suggest below that the emphasis on Italian humanism is misplaced, Reynolds's engagement with the most explicitly secular dimension of the model, its text, signals a new appreciation of the ways in which sacred and secular commingle in this genre.

8. Howard Mayer Brown, "Music and Ritual at Charles the Bold's Court: The Function of Liturgical Music by Busnoys and His Contemporaries," in Paula Higgins, ed., *Antoine Busnoys: Method, Meaning, and Context in Late Medieval Music* (Oxford: Clarendon, 1999), 53–70, at 64.

9. Figure 1.1 flings its net well beyond the repertory on which this essay concentrates, encompassing masses based on songs in English, Italian, and German as well as French, and pushing the temporal span up to the death of Duke Charles the Bold in 1477 (a date suggested by the cultural milieu explored here). Omitted from the table are the many masses based on *L'homme armé* and masses based on monophonic dance tunes. There are of course a number of song masses composed before 1477 but surviving only in later sources; these would include, for example, the *Missa Je suis en la mer* by Faugues, the *Missa Entrepris suis* by Vincenet, and the anonymous *Missa Nos amis*, all in ModE M.1.13, copied ca. 1481.

10. For biographical summaries, see the relevant entries and bibliographies in *New Grove II*.

11. Reinhard Strohm notes the seminal role of English composers in the creation of chanson masses in *The Rise of European Music, 1380–1500* (Cambridge: Cambridge University Press, 1993), 413; see also his Table 5 on 430–31. On English and Burgundian contact during this period, see Richard Vaughan, *Philip the Good: The Apogee of Burgundy* (Harlow: Longmans, 1970), esp. 1–28 and 98–126.

12. Richard Taruskin first suggested this origin for chanson masses in Antoine Busnoys, *Collected Works: The Latin-Texted Works*, 2 vols., ed. Richard Taruskin, Masters and Monuments of the Renaissance 5 (New York: Broude Trust, 1990), commentary, 95–100.

13. Precisely why the Trent codices were copied is unknown; the capability for performance of complex polyphony at Trent Cathedral during this period appears to have been limited. Much of the repertory transmitted in these sources may have been obtained through the Austrian court chapels in Innsbruck and Wiener Neustadt; see Adelyn Peck Leverett, "Song Masses in the Trent Codices: The Austrian Connection," *Early Music History* 14 (1995): 205–56.

14. Christopher Reynolds, "The Origins of San Pietro B 80 and the Development of a Roman Sacred Repertory," *Early Music History* 1 (1981): 257–304.
15. The 1451 marriage was proposed by Alejandro Enrique Planchart in "Fifteenth-Century Masses: Notes on Performance and Chronology," *Studi musicali* 10 (1981): 5–29, at 5; the 1452 event by David Fallows, *Dufay* (London: J. M. Dent, 1982), 70.
16. See Robert J. Snow, "The Mass-Motet Cycle: A Mid-Fifteenth-Century Experiment," in Gustave Reese and Robert J. Snow, eds., *Essays in Musicology in Honor of Dragan Plamenac on His 70th Birthday* (Pittsburgh: University of Pittsburgh Press, 1969), 301–20.
17. Leverett, "Song Masses," 223–26.
18. On Origen and the early commentaries, see E. Ann Matter, *The Voice of My Beloved: The Song of Songs in Western Medieval Christianity* (Philadelphia: University of Pennsylvania Press, 1990), 20–122.
19. This important point is made by Rachel Fulton in "Mimetic Devotion, Marian Exegesis, and the Historical Sense of the Song of Songs," *Viator: Medieval and Renaissance Studies* 27 (1996): 85–116.
20. Johannes Beumer, "Die marianische Deutung des Hohen Liedes in der Frühscholastik," *Zeitschrift für Katholische Theologie* 76 (1954): 411–39. See also Matter, *The Voice of My Beloved,* 153–58.
21. Regarding the profound impact of the Song of Songs on medieval culture, see first and foremost Matter, *The Voice of My Beloved,* esp. 151–77, and Ann W. Astell, *The Song of Songs in the Middle Ages* (Ithaca: Cornell University Press, 1990). See also Marina Warner, *Alone of All Her Sex: The Myth and Cult of the Virgin Mary* (New York: Knopf, 1976), 121–33; Peter Dronke, *Medieval Latin and the Rise of the European Love-Lyric,* 2 vols. (Oxford: Clarendon, 1965–66), vol. 1; and Jean Leclercq, *Monks and Love in Twelfth-Century France: Psycho-Historical Essays* (Oxford: Clarendon, 1979), esp. 27–61 and 121–29.
22. This is Le Mans B. M. 173, fols. 33v–110r. See Cedric E. Pickford, ed., *The Song of Songs: A Twelfth-Century French Version* (London, New York, and Toronto: Oxford University Press for the University of Hull, 1974).
23. Literary theorists have been particularly attuned to this acute intertextuality: see in particular Paul Zumthor, *Langue et techniques poétiques à l'époque romaine (XIe–XIIIe siècles)* (Paris: C. Klincksieck, 1963), and Giovanni Pozzi, "Codici, stereotipi, topoi e fonte letterarie," in *Intorno al codice: atti del III Convegno della Associazione italiana di studi semiotici, Pavia, 26–27 settembre, 1975,* Pubblicazioni della Facoltà di lettere e filosofia dell'Università di Pavia 22 (Florence: La nuova Italia, 1976), 37–76.
24. Peter Dronke, "The Song of Songs and Medieval Love-Lyric," in W. Lourdaux and D. Verhelst, eds., *The Bible and Medieval Culture,* Mediaevalia Lovaniensia I/7 (Leuven: Leuven University Press, 1979), 236–62.
25. On the scriptural and specifically Marian dimensions of the *Roman de la Rose,* see John V. Fleming, *The Roman de la Rose: A Study in Allegory and Iconography* (Princeton: Princteon University Press, 1969), esp. 54–184. On Dante's Mariology, see Warner, *Alone of All Her Sex,* 160–74, and Jaroslav Pelikan, *Mary through the Centuries: Her Place in the History of Culture* (New Haven and London: Yale University Press, 1996), 139–51.
26. Discussed in Dronke, *Medieval Latin and the Rise of the European Love-Lyric,* 1:59–63, whence this translation is drawn.
27. This transformation was noted by Émile Mâle more than a century ago. See his *Religious Art in France, The Thirteenth Century: A Study of Medieval Iconography and Its Sources,* trans. Marthiel Mathews, Bollingen Series 90/2 (Princeton: Princeton University Press, 1984), 233–66, esp. 235–39. The original publication was *L'Art religieux du XIIIe siècle en France: Étude sur l'iconographie du moyen âge et sur ses sources d'inspiration* (Paris: Armand Colin, 1898); the translation is from the 9th ed., 1958. Other thought-provoking studies concerned with the representation of the Virgin in this period are Penny Schine Gold, *The Lady and the Virgin: Image, Attitude, and Experience in Twelfth-Century France* (Chicago: University of Chicago Press, 1985), 43–75; and Michael Camille, *The Gothic Idol: Ideology and Image-Making in Medieval Art* (Cambridge: Cambridge University Press, 1989), 220–41 and 298–316.
28. Gold, *The Lady and the Virgin,* Plates 4 and 6, and 8 and 9, discussed on 73.
29. On the *Maria lactans* theme, see Warner, *Alone of All Her Sex,* 192–205; on the *Mater dolorosa* see ibid., 206–23.

30. John Stevens, *Words and Music in the Middle Ages: Song, Narrative, Dance and Drama, 1050–1350* (Cambridge and New York: Cambridge University Press, 1986), 395.
31. Matter, *The Voice of My Beloved*, 189.
32. Sylvia Huot, *Allegorical Play in the Old French Motet: The Sacred and the Profane in Thirteenth-Century Polyphony* (Stanford: Stanford University Press, 1997), esp. 85–127.
33. Kevin Brownlee, "Machaut's Motet 15 and the *Roman de la Rose:* The Literary Context of *Amours qui a les pouoir/Faus Samblant m'a deceü/Vidi Dominum*," *Early Music History* 10 (1991): 1–14; Margaret Bent, "Deception, Exegesis and Sounding Number in Machaut's Motet 15," *Early Music History* 10 (1991): 15–27; Huot, *Allegorical Play*, 193–94.
34. For a useful summary of the application of analogy prior to the fifteenth century, see Michael Randall, *Building Resemblance: Analogical Imagery in the Early French Renaissance* (Baltimore and London: The Johns Hopkins University Press, 1996), 9 and 41–42.
35. I borrow this phrase, "genre as trope," from Matter's final chapter in *The Voice of My Beloved*, having to do with the use of the Song of Songs in vernacular literature.
36. This is not to deny the many symbolic details that sanctify the space, as noted by Barbara G. Lane, *The Altar and the Altarpiece: Sacramental Themes in Early Netherlandish Painting* (New York: Harper and Row, 1984), 45–47.
37. Sandro Lombardi reviews the evidence supporting this correspondence in *Jean Fouquet* (Florence: Libreria Editrice Salimbeni, 1983), 125–35. The cultural historian Johan Huizinga, horrified by the possibility, saw in this Madonna "an air of decadent impiety . . . a flavour of blasphemous boldness about the whole unsurpassed by any artist of the Renaissance." See 143 in Johan Huizinga, *The Waning of the Middle Ages: A Study of the Forms of Life, Thought and Art in France and the Netherlands in the XIVth and XVth Centuries*, rev. and trans. under the author's supervision by F[rederik] Hopman (London: E. Arnold, 1924), from *Herfsttij der Middeleeuwen: Studie over levens- en gedachtenvormen der veertiende en vijftiende eeuw in Frankrijk en de Nederlanden* (Haarlem: H. D. Tjeenk Willink, 1919).
38. Huizinga, *The Waning of the Middle Ages*, 136–59; masses based on songs are mentioned on 141.
39. The standard work on Gerson remains James L. Connolly, *John Gerson: Reformer and Mystic*, Université de Louvain, Recueil de travaux publiés par les membres des Conférences d'histoire et de philologie II/12 (Leuven: Librairie Universitaire; London and St. Louis: B. Herder, 1928). Gerson did not remain in the good favor of the Burgundians after 1408, when he spoke out against the murder of Louis, Duke of Orléans, by the Burgundian duke John the Fearless in November 1407.
40. For a useful summary of Gerson's theology, see Randall, *Building Resemblance*, 15–21. Gerson was not, of course, the first to formulate a theology that challenged scholastic realism; in the fourteenth century the nominalist critique of William of Ockham and the mystical theologies of Meister Eckhart and others weakened the scholastic edifice. But Gerson initiated an educational campaign that nourished personal piety from within rather than in opposition to mainstream church teaching, and made a deep and lasting impression on devotional attitudes in northern Europe. His use of the vernacular in many of his writings, exceptional for a theologian at this time, further attests to his purpose and explains his influence.
41. See Jean Gerson, *Œuvres complètes*, ed. Palémon Glorieux (Paris and New York: Desclée, 1960–73), vol. 7: *L'Œuvre française*, 29–30; for a translation see Jean Gerson, *Early Works*, trans. Brian Patrick McGuire, Classics of Western Spirituality 92 (New York: Paulist, 1998), 92–93.
42. Gerson, *Œuvres complètes*, 7:144.
43. For a detailed study of this poetry, see Paul Zumthor, *Le Masque et la lumière: La Poétique des grands rhétoriqueurs* (Paris: Seuil, 1978).
44. Gerson, *Œuvres complètes*, 7:152.
45. On Molinet, see the classic study by Noël Dupire, *Jean Molinet: La Vie—Les Œuvres* (Paris: E. Droz, 1932), and Jean Devaux, *Jean Molinet: Indiciaire Bourguignon*, Bibliothèque du XVe siècle 55 (Paris: Honoré Champion, 1996).
46. For the complete text of the poem, see Jean Molinet, *Les faictz et dictz de Jean Molinet*, ed. Noël Dupire (Paris: Société des anciens textes français, 1936–39), 2:531–35. Dupire treats this poem as an example of parody; see his *Jean Molinet: La Vie—Les Œuvres*, 117.

47. On the idea of register in medieval poetry, see Paul Zumthor, "Style and Expressive Register in Medieval Poetry," in Seymour Chatman, ed. and (in part) trans., *Literary Style: A Symposium* (London and New York: Oxford University Press, 1971), 263–91.
48. Sixten Ringbom, *Icon to Narrative: The Rise of the Dramatic Close-Up in Fifteenth-Century Devotional Painting,* Acta academiae aboensis Ser. A, 31/2 (Åbo: Åbo Akademi, 1965), 48–49.
49. The fifth line of this eighth verse, "La plus des plus," is likewise the incipit of a chanson text; see Fallows, *A Catalogue of Polyphonic Songs,* 235. Molinet's interest in music is manifest throughout his writings; see Carol MacClintock, "Molinet, Music, and Medieval Rhetoric," *Musica Disciplina* 13 (1959): 109–21, and Albert van der Linden, "La Musique dans les Chroniques de Jean Molinet," in *Mélanges Ernest Closson: Recueil d'articles musicologiques offert à Ernest Closson à l'occasion de son soixante-quinzième anniversaire* (Brussels: Société belge de musicologie, 1948), 166–80. On his use of chanson incipits in particular, see Françoise Ferrand, "Le Grand Rhétoriqueur Jean Molinet et la chanson polyphonique à le Cour des Ducs de Bourgogne," Danielle Buschinger and André Crépin, eds., in *Musique, littérature et société au moyen âge: Actes du colloque, 24–29 Mars 1980* (Paris: Honoré Champion, for Université de Picardie, Centre d'études médiévales, 1980), 395–407, and Clemens Goldberg, "Militat omnis amans: Zitat und Zitieren in Molinets 'Le débat du viel Gendarme et du viel amoureux' und Ockeghems Chanson 'L'autre d'antan,' " *Musikforschung* 42 (1989): 341–49.
50. For the complete poem, see Molinet, *Les faictz et dictz,* 2: 468–75.
51. Despite Huizinga's dim view of such playfulness in northern court culture of the time, *The Waning of the Middle* Ages remains the most thorough survey of this aspect; see especially 182–94 and 222–96.
52. "[E]t convertir soubz mes rudes meules les vicieux au vertueux/le corporel en spirituel/la mondanité en divinité", quoted in Randall, *Building Resemblance,* 23, who offers a thoughtful analysis of this much-maligned moralization on 13–39.
53. Conrad von Zabern, *De modo bene cantandi* (Mainz: Peter Schöffer, 1474). Conrad's heated condemnation is worth quoting in full: "Item etiam sic est cantandum devotionaliter, ut nulla melodia, quae a devotis sanctis patribus nobis non est tradita, sed a diaboli ministris introducta, ut infra dicetur, inter divinae laudis carmina umquam cantetur; huiusmodo enim melodiae adulterinae penitus sunt reiciendae a divino officio. Et tamen heu in plerisque ecclesiis quasi cotidie sunt in usu. Exemplariter loquar, ut intelligar: Nonnulli scolarium rectores placere nescio cui cupientes, sed haud dubium diabolo per hoc servientes, etiamsi nescii, quorundam mundialium carminum melodias sumpserunt et illas super his, quae de potioribus sunt inter divinae laudis carmina, hoc est super hymnum angelicum *Gloria in* excelsis et super Symbolum Nicaenum ac super *Sanctus* et *Agnus Dei,* ut poterant, aptarunt haec sub eisdem mundialibus melodiis cantando dimissis devotis sanctorum patrum melodiis nobis praescriptis. Quae mundialium christifideles, ut sciens scio, scandalisent, sed etiam multos praesertim iuvenes vel carnales homines plus de domo choreae quam de regno caelorum cogitare faciunt in devotionis impedimentum non modicum, nimirum quia huiusmodi melodias vel eis similes in domo choreae saepe audierunt." See Conrad von Zabern, *Die Musiktraktate Conrads von Zabern,* ed. Karl-Werner Gümpel, Akademie der Wissenschaften und der Literatur, Abhandlungen des geistes- und sozialwissenschaftlichen Klasse, Jahrgang 1956, 4 (Wiesbaden: F. Steiner for Akademie der Wissenschaften und der Literature, Mainz, 1956), 271. I am indebted to Dr. Bonnie Blackburn for bringing Conrad's diatribe to my attention.
54. Craig Wright, "Performance Practices at the Cathedral of Cambrai 1475–1550," *Musical Quarterly* 64 (1978), 295–328, at 299.
55. Edith Weber, *Le Concile de Trente et la musique: De la Réforme à la Contre-Réforme* (Paris: Honoré Champion, 1982), esp. 85–90 and 137–43.
56. August Wilhelm Ambros, *Geschichte der Musik III* (Leipzig: F. E. C. Leuckart, 1868; 3rd ed., rev. Otto Kade, 1893), 46. On Ambros's definitive role in the modern historical construction of the Mass genre, see Andrew Kirkman, "The Invention of the Cyclic Mass," *JAMS* 54 (2001): 1–47. I am grateful to Dr. Kirkman for sharing his essay in advance of its publication.
57. Peter Wagner, *Geschichte der Messe,* Kleine Handbücher der Musikgeschichte nach Gattungen 11/1 (Leipzig: Breitkopf & Härtel, 1913), 64.
58. See the seminal study by Howard Mayer Brown, "Emulation, Competition, and Homage: Imitation and Theories of Imitation in the Renaissance," *JAMS* 35 (1982): 1–48, esp. 44–48

concerning the relevance of *imitatio* to the mass repertory; for further development of the idea in relation to masses based on secular polyphonic models, see Burkholder, "Johannes Martini and the Imitation Mass," with critiques by Leeman L. Perkins, "Communication," *JAMS* 40 (1987): 130–34; and Reinhard Strohm, "Communication," *JAMS* 40 (1987): 576–79. Honey Meconi argues persuasively against assuming a direct and causal link between rhetorical *imitatio* and borrowing procedures in music of the fifteenth and sixteenth century in "Does *Imitatio* Exist?" *Journal of Musicology* 12 (1994): 152–78.

59. The key to the analysis of Josquin's *Missa di dadi* offered by Long in "Symbol and Ritual" is a fine example of what the consideration of "inaudible polytextuality" can illuminate.

2
Ockeghem and Intertextuality: A Composer Interprets Himself

MURRAY STEIB

As polyphonic chansons became an important source for mass cantus firmi in the mid-fifteenth century, composers also had the new option of using polyphonic quotations. Until recently, however, these polyphonic borrowings were seldom studied. My work on this subject has shown that the number of masses incorporating polyphonic quotations is far greater than previously suspected and spans several generations, from the earliest cantus firmus masses of Dufay and Bedyngham in the mid-fifteenth century to those of Josquin and his contemporaries at the end of the century.[1] Indeed, virtually every mass based on a polyphonic model uses polyphonic quotations to some degree. Furthermore, I have shown that composers incorporated borrowed material in three ways. The most common method is the use of extended, literal quotations, in which entire phrases of the model are quoted exactly, or with only a few added ornamental notes; the second is the use of paraphrased quotations, in which the borrowed material is subject to a great deal of ornamentation and is often paraphrased beyond immediate recognition; and the third is the combination of both literal and paraphrased quotations together. In all three categories, composers also make other polyphonic allusions to their models by expanding and elaborating points of imitation from the model, creating new points of imitation where none existed before, and occasionally combining phrases that were not originally associated with one another. One of the most important observations to be made concerning these three methods of borrowing is that composers generally did not use several methods during their careers, but rather employed only one of the methods in all of their masses. This consistency makes the study of polyphonic borrowings an extremely valuable tool in style criticism.

Ockeghem's masses are unrivaled in both quality and invention and continue to merit our closest attention. He uses monophonic as well as polyphonic models and models written by himself as well as others; some of his masses do not use a cantus firmus but do allude to other works, while other masses appear to be free of any borrowed material whatsoever. He enjoyed setting virtuosic technical hurdles in his masses: *Missa prolationem* is built around a series of double mensuration canons, and *Missa cuiusvis toni* can be sung in any mode by reading the music in different clef combinations. Despite using such techniques, his music is never dry and academic, but always fresh and sublime. Moreover, he was very prolific for his generation, composing ten complete masses, three partial or fragmentary masses, one Requiem, and one individual mass movement.[2] Given his importance, stature, and oeuvre, it is surprising that his borrowing techniques—especially his polyphonic borrowing techniques—have not been studied thoroughly. There are several reasons for this. The study of polyphonic borrowing did not become an important field of inquiry until the 1980s, and even since then, very few people have realized how important Ockeghem is in this field.[3] Furthermore, his masses not based on polyphonic models—such as *Missa Mi mi*—are so spectacularly attention-grabbing that his other masses tend to be overlooked.

Studies of Ockeghem that examine matters of style have focused primarily on his complex treatment of cantus firmi but neglected polyphonic borrowing. This has distorted our view of Ockeghem's masses and their overall relatedness to their models. By studying his polyphonic quotations, however, we can obtain a more complete picture of how Ockeghem borrowed and gain insight into how he viewed his models. This is especially important considering that four of his masses are based on polyphonic models and borrow simultaneously from more than one voice of their model. Another two, although not based on polyphonic models, nevertheless allude to them (see Figure 2.1).

For the purposes of this study, I would like to draw a distinction between these two categories. A mass based on a polyphonic model is one that uses a cantus firmus as its structural basis, even if only for a single movement, as in *Missa Au travail suis;* a mass that alludes to a polyphonic model does not use that model as a cantus firmus but occasionally makes either monophonic or polyphonic reference to it. A mass (or a motet or chanson) can

Masses Based on Polyphonic Models	Composer of the Model
Missa Fors seulement	Ockeghem
Missa Ma maistresse	Ockeghem
Missa De plus en plus	Binchois
Missa Au travail suis	Ockeghem or Barbingant
Masses Alluding to Polyphonic Models	**Composer of the Model**
Missa Mi mi	Ockeghem, *Presque transi*
Missa L'homme armé	Morton, *Il sera pour vous/L'homme armé* and Ockeghem, *L'aultre d'antan*

Fig. 2.1 Ockeghem's masses based on or alluding to polyphonic models.

be based on one model yet still allude to another, or it can allude to one or more models even if it is not based on one. As an example of the latter, Jeffrey Dean has shown that *Intemerata Dei mater* alludes to a number of Ockeghem's own pieces, including *Missa Mi mi* and the chansons *Presque transi* and *Fors seulement l'actente*;[4] as an example of the former, Christopher Reynolds has demonstrated that Caron's *Missa Accueilly m'a la belle* alludes to Binchois's *De plus en plus*.[5]

Howard Mayer Brown once suggested that in order to identify what is individual about a composer, we ought to concentrate on broad issues such as compositional strategies, habits, and practices rather than focusing on details that might just as well have been the work of an editor, publisher, or scribe as of the composer himself.[6] Many such fundamental questions concerning Ockeghem's borrowing procedures—involving both cantus firmus and polyphonic quotations—have yet to be addressed. Are his polyphonic quotations paraphrased, literal, or some combination of the two? Do they resemble or differ from his method of using a cantus firmus? Is he consistent in his use of polyphonic quotations? Are quotations used frequently throughout a mass or only occasionally? Do they involve the entire model or only small portions of it? Are some sections of a mass less likely to contain quotations than others, just as some sections often omit the cantus firmus? By examining these and similar questions, I hope to refine our understanding of his personal style, which in turn can help answer broader questions about musical borrowing. Is Ockeghem using his own model or someone else's, as in the case of *Missa Au travail suis?* What can his borrowing procedures tell us about the model for *Missa L'homme armé?* Are there chronological implications for his borrowing practices? Can his borrowing practices help us to situate his masses with one another, at least relatively?

I will begin by examining Ockeghem's masses definitely based on his own models (*Missae Fors seulement* and *Ma maistresse*), followed by his one mass definitely based on someone else's model (*Missa De plus en plus*, based on a chanson by Binchois), and then the mass based on a model whose authorship is in question (*Missa Au travail suis*, based on a model variously attributed to Ockeghem and Barbingant). Finally, I will analyze his masses that contain allusions to other works without using a cantus firmus (*Missae Mi mi* and *L'homme armé*).

Before turning to the works themselves, a few words about terminology are in order. We use a variety of terms to describe the process of musical borrowing—elaboration, *imitatio*, imitation, paraphrase, parody, and so forth—but many of these terms have extramusical associations that may obfuscate rather than illuminate the relationships they attempt to describe. I have chosen the more neutral term intertextuality to describe the relatedness between pieces. One of the virtues of this term is that it can be used to describe different kinds of relationships (melodic, contrapuntal, structural, and textual) and different modes of relatedness (literal, ornamented, or paraphrased). Although the term is of relatively recent use in musicological studies, it has been used in literary criticism for much longer.[7]

Ockeghem's Masses Securely Based on His Own Models

Ockeghem's *Missa Fors seulement* is based on his own rondeau cinquain *Fors seulement l'actente* and consists of only three movements: Kyrie, Gloria, and Credo. The cantus firmus is derived from the tenor and discantus of the chanson and is heard at least once in each voice (see Figure 2.2).[8] After the Kyrie, phrases are occasionally repeated or omitted, and twice they migrate from one voice to another. This somewhat unusual approach to the cantus firmus in no way diminishes the relationship between mass and model. The cantus firmus is stated literally, with few changes of either pitch or rhythm. For example, if we compare the last two phrases of the cantus firmus in the Kyrie with the model, we see very few modifications, and those few that do occur are relatively minor: combining repeated notes into a single note, probably for textual reasons (chanson, mm. 55, 57, and 61) and adding a single passing tone (m. 72); see Figure 2.3.[9]

Missa Fors seulement contains many other references to its model besides the cantus firmus. For example, Ockeghem treats several phrases of the model canonically, as at the beginning of the Kyrie, where the discantus and contratenor both state phrase A of the chanson discantus, in canon at the fifth, a breve apart (see Figure 2.4).[10] In the chanson, the contratenor outlines a diminished triad in the opening measures, rising from A to E-flat (in most versions); it is interesting to note that in the quotation of

Movement	Phrase of Cantus Firmus	Original Voice of C.F. in Model	Placement of C.F. in Mass
Kyrie			
Kyrie	A	Discantus	Discantus
	BC	Tenor	Tenor
Christe	D	Tenor	Tenor
Kyrie	E	Discantus	Tenor
	FG	Tenor	Contratenor
Gloria			
Et in terra	B	Tenor	Vagans
	A	Discantus	Discantus
	C	Tenor	Vagans & Tenor
	D	Tenor	Discantus
	D	Discantus	Tenor
Qui tollis	EF	Discantus	Tenor
	G	Tenor	Tenor
	G	Discantus	Vagans
Credo			
Patrem	BCDFG	Tenor	Tenor I
Et resurrexit	FF	Discantus	Tenor I
	G	Tenor	Tenor I

Fig. 2.2 The layout of the cantus firmus in Ockeghem's *Missa Fors seulement.*

the contratenor in this example, Ockeghem not only transposed the line but also eliminated the accidental in m. 5, thus avoiding both a melodic and harmonic diminished triad in the mass. Another canonic example occurs in mm. 106–11 of the Gloria, where the opening of phrase G of the tenor of the model is treated canonically, at the fifth and a breve apart, in the tenor and bassus.

Another way that Ockeghem manipulates borrowed material is by combining phrases that were not originally associated with one another in the model. For example, at the beginning of the Gloria, he combines phrase A of the discantus with phrase B of the tenor (stated in the contratenor and vagans, respectively), both transposed down a fifth.[11] Although these two phrases begin identically, making this passage look like another canon, they end differently, making this passage somewhat different from the previous two.

Fig. 2.3 A comparison of the superius of *Fors seulement l'actente* and its use as the cantus firmus in the Kyrie of *Missa Fors seulement*.

Finally, Ockeghem also uses three polyphonic quotations in this mass, one in each movement. None of these quotations borrows from the same section of the chanson: the first comes from the opening of the chanson, the second from the middle, and the third from the end. The last of these quotations encompasses an entire phrase of the model (see Figure 2.5), whereas the others encompass less than a complete phrase (see Figure 2.4). In Figure 2.5, the cantus firmus is stated in the bassus and contains no changes to the chanson discantus. In addition, the tenor of the chanson is quoted in the tenor of the mass; it too is very close to the original, with only a few passing tones added (mm. 190, 192, and 195) and a breve divided into

Fig. 2.4 Ockeghem, *Missa Fors seulement,* Kyrie, mm. 1–7.

a semibreve and a rest (mm. 196–97). The other polyphonic quotations in this mass contain a similarly small amount of modification, as in Figure 2.4. Here the contratenor states phrase A of the discantus transposed down a fifth, and the first bassus has phrase A of the contratenor, also transposed down a fifth. In this example, both voices involved in the quotation have combined repeated semibreves into a single breve in m. 5; the first bassus has divided what was a long F into a breve D and a breve F in mm. 3–4; and it has also omitted a dissonant passing tone in m. 6. Despite these alterations, this quotation is still very close to the original and is clearly recognizable.

Ockeghem's *Missa Ma maistresse* consists of only two movements (Kyrie and Gloria) and is based on his own bergerette. The cantus firmus is derived from the tenor of the chanson and placed in the bassus of the mass in the Kyrie, and derived from the superius and placed in the contratenor in the Gloria. The cantus firmus is stated literally, with few changes in either pitch or rhythm, as it was in *Missa Fors seulement.* In *Missa Ma maistresse,* Ockeghem again has many references to the model in addition to the cantus firmus. He treats several phrases of the model canonically, as he did in *Missa Fors seulement;* for example, the opening of the Gloria has phrase A of the superius in canon at the octave, a breve apart, in the contratenor and bassus.

Missa Ma maistresse has five polyphonic quotations in its two movements, which is more than there were in *Missa Fors seulement.* Most of the quotations—four of the five—are in the Kyrie, and except for the first two

Fig. 2.5 Ockeghem, *Missa Fors seulement,* Credo, mm. 188–97.

Fig. 2.6 Ockeghem, *Missa Ma maistresse,* Kyrie, mm. 10–14.

at the very beginning of the Kyrie, which both quote the opening two measures of the chanson, they quote from different sections of the model. In this regard, *Missa Ma maistresse* is similar to *Missa Fors seulement*. Figure 2.6 is typical of the way Ockeghem incorporates polyphonic quotations in this mass: the cantus firmus is in the bassus and contains only one change to the model, the addition of a rest in m. 12;[12] the quotation of the superius begins in the tenor in m. 10 but migrates to the contratenor in m. 13, and is also very close to the original, with only a small amount of ornamentation in m. 12. The relatively brief quotation of the contratenor in m. 12 in the superius may not be a quotation at all but simply a fortuitous resemblance; however, it is worth noting that the pitches, rhythms, and alignment are virtually identical to the original. The other polyphonic quotations in this mass are similar to Figure 2.6. The quotation in mm. 28–33 of the Kyrie, for example, is virtually identical to this example in length, degree of relatedness, and in that the quotation of the superius begins in the tenor and migrates to the contratenor. The other polyphonic quotations in this mass are somewhat shorter and contain no migration, but like Figure 2.6, they quote the chanson literally with little or no ornamentation.

Ockeghem's Mass Based on a Model by Another Composer

Ockeghem's *Missa De plus en plus* is based on Binchois's rondeau cinquain. It was probably composed earlier than any of the other masses examined so far,[13] and in several respects it is vastly different. The cantus firmus in his

Missa De plus en plus is paraphrased, which contrasts with the cantus firmi in *Missa Fors seulement, Missa Ma maistresse,* and *Missa Au travail suis.* Figure 2.7, illustrating Ockeghem's approach to paraphrase in the cantus firmus, compares the tenor of the chanson with the cantus firmus from the Kyrie. Here Ockeghem goes beyond merely adding the occasional ornamental note, and in many places he has so reshaped the cantus firmus with melodic extensions and excursus that the original line is barely recognizable. Some of the cantus firmus in this mass is very close to the original in terms of pitch content and rhythmic structure, but most of it is melodically and rhythmically paraphrased. The most common type of change that he makes in the cantus firmus is decorating individual long notes (this occurs in all phrases). In these instances, he often repeats a note from the model immediately after a melodic digression and before moving on to the next note from the model (see mm. 10, 18, 20, 30, and 35).[14] Another common change that he makes is filling in melodic skips with passing tones.

Missa De plus en plus contains fifteen polyphonic quotations, far more than either *Missae Fors seulement* or *Ma maistresse.* As in the other masses already examined, this mass quotes from many different parts of the model, although there is some preference given to the opening of the chanson.[15] The polyphonic quotations in this mass, like the cantus firmus, vary from literal to heavily paraphrased and show no preference for either method. In this respect, the polyphonic quotations in this mass are different from his other masses, in which he does not paraphrase borrowed material but rather states it literally or with a modicum of ornamentation. Furthermore, they are brief and most frequently quote only about half of a phrase of the model or less. Such brief, paraphrased quotations were common during the second half of the fifteenth century and are also found in the masses of Josquin.[16]

A typical example of how Ockeghem incorporates a polyphonic quotation with a small amount of ornamentation may be seen in Figure 2.8. In this passage, the tenor states the cantus firmus—taken from the second half of phrase E, the last phrase of the model—literally, and the superius quotes the superius of the model with only a single ornamental note added (in m. 128) and one note omitted (in m. 127). In the last four measures of this example, the contratenor quotes the contratenor of the model. About half of the quoted notes in the superius and contratenor are in the same alignment with the tenor as they were in the chanson, as indicated by the arrows. Although this is only a lightly ornamented quotation, the difference between it and earlier examples is readily apparent; compare Figure 2.8 with Figures 2.4, 2.5, 2.6, and 2.7.

Ockeghem also incorporates heavily paraphrased polyphonic quotations in this mass, as in Figure 2.9b, taken from the end of the Kyrie of *Missa De plus en plus.* Here both the cantus firmus and the quotation of the superius of

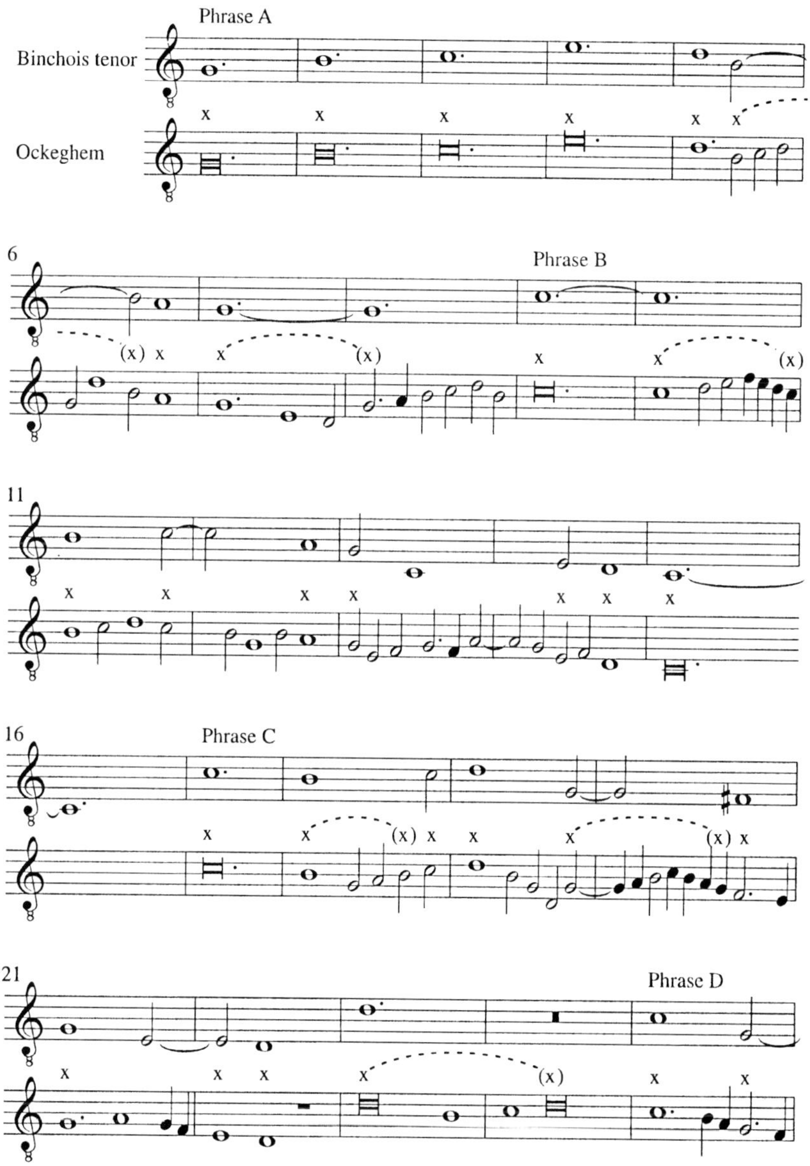

Fig. 2.7 A comparison of the tenor of Binchois's *De plus en plus* and the cantus firmus in the Kyrie of Ockeghem's *Missa De plus en plus*.

Fig. 2.7 *(cont.)*.

Fig. 2.8 Ockeghem, *Missa De plus en plus,* Gloria, mm. 127–32.

the model are paraphrased. Ockeghem clarifies this paraphrased passage in several ways. He occasionally repeats a note from the model after a melodic digression and before moving on the next note from the model, as in m. 42 of the tenor. He maintains the original alignment for about half of the passage, as indicated by the arrows, and even in those places where he has not kept the exact alignment between superius and tenor, they are usually close. For example, the first pitch quoted from the superius (the G in m. 40) originally aligned with the G in the tenor in m. 41, and the second quoted pitch in the superius (the E in m. 41), instead of sounding simultaneously with the G in the tenor, originally sounded just after the G.

Ockeghem's Mass Based on a Model of Uncertain Attribution

Ockeghem's *Missa Au travail suis* is based on a rondeau quatrain of uncertain authorship. The chanson exists in three sources: in ParisBN 57 it is attributed to Ockeghem, in DijBM 517 it is attributed to Barbingant, and in WolfA 287 it is anonymous. Current scholarly opinion is still divided on the issue of authorship. The article on Ockeghem in *New Grove* gives the work to Ockeghem;[17] David Fallows gives the work to Barbingant but admits that there is some ambiguity involved;[18] and although the newly completed collected works of Ockeghem lists the chanson under "Doubtful Works," Richard Wexler admits, "There is nothing in the style of the chanson to preclude its having been written by either composer."[19]

Fig. 2.9 (a) Binchois, *De plus en plus,* mm. 1–8; (b) Ockeghem, *Missa De plus en plus,* Kyrie, mm. 40–48.

Missa Au travail suis is not a cantus firmus mass in the traditional sense. The Kyrie uses the tenor of the chanson as a cantus firmus, but the Gloria, Credo, Sanctus, and Agnus do not use a cantus firmus at all. In the Kyrie of *Missa Au travail suis,* the chanson tenor is stated once in its entirety as the cantus firmus, transposed down a fifth. The opening point of imitation in the chanson is expanded in the Kyrie to include all four voices. The last four movements open with a brief quotation of the opening phrase of the chanson tenor, which resembles a motto rather than a cantus firmus. These quotations vary in length from five to seven measures, and the opening three or four notes are imitated in at least one other voice. As Fitch has pointed out, some of the sectional endings of these last four movements resemble the closing measures of the chanson.[20] Thus there is an allusion to the model at both the opening and the closing of each of these last four movements.

There are only two polyphonic quotations in *Missa Au travail suis.* This small number is typical of his other masses based on models securely attributed to Ockeghem. Furthermore, although he opens all of the movements with melodic material based on the opening of the chanson, the polyphonic quotations per se in this mass quote different sections of the model. One example of a polyphonic quotation comes from in the middle of the second Kyrie and is very similar to the polyphonic quotations in Ockeghem's *Missae Fors seulement* and *Ma maistresse:* it is short and literal (see Figure 2.10). In this example, the cantus firmus is transposed down a fifth (as it is throughout the movement), but the quotation of the superius is transposed down a third.[21] Despite the different levels of transposition in this quotation, Ockeghem has maintained the same rhythmic structure and the same vertical alignment of pitches as in the chanson.

Fig. 2.10 Ockeghem, *Missa Au travail suis,* Kyrie, mm. 33–34.

Fig. 2.11 (a) Ockeghem, *Ma maistresse,* mm. 1–5; (b) *Au travail suis,* mm. 16–20; (c) Ockeghem, *Missa Au travail suis,* Kyrie, mm. 26–30.

The other polyphonic quotation in *Missa Au travail suis*—from the opening of Kyrie II (mm. 26–30)—is interesting for several reasons. The section from the chanson that is being quoted is itself a quotation from another chanson, Ockeghem's *Ma maistresse,* and thus is a quotation of a quotation. The impetus behind the quotation is clear: the text in *Au travail suis* at this point is "Ma maistresse," and the musical quotation takes place at the very time the two chansons share the same text. Evidently the composer wanted the musical allusion as well as the textual one. The only change in the quotation occurs at the beginning, where the first note in *Ma maistresse*—a long—is divided into a dotted breve and a semibreve in *Au travail suis* (see Figures 2.11a and 2.11b). At this point in both chansons, the superius is imitated in the tenor, but in slightly different ways. In *Ma maistresse* the two voices are a fifth apart, and although the two voices begin simultaneously, the superius is the *dux* and the tenor is the *comes.* In *Au travail suis,* on the other hand, the two voices are in octaves; the tenor begins first and is the *dux,* while the superius enters a semibreve later and is the *comes.* In *Missa Au travail suis,* the cantus firmus is clearly derived from *Au travail suis* and is transposed down a fifth (see Figure 2.11c). As in the two chansons, the mass at this point has another voice imitating the tenor, in this case the altus. Here, however, Ockeghem seems to be quoting not *Au travail suis* but rather *Ma maistresse.* The altus is a fifth higher than the tenor, the altus is the *dux* while the tenor is the *comes,* and the vertical alignment is much closer to *Ma maistresse* than to *Au travail suis.* This quotation is similar to Ockeghem's other quotations examined so far in that it is literal and comprises less than a complete phrase from the model.

Missa Mi mi and *Presque transi*

There is still some debate concerning Ockeghem's *Missa Mi mi* and its possible relatedness with his own bergerette *Presque transi.* A connection between these two pieces was first made by Haruyo Miyazaki and subsequently expanded upon by Wolfgang Thein, Rebecca Stewart, Fabrice Fitch, and Jaap van Benthem.[22] Recently, however, Martin Picker argued that Ockeghem did not consciously quote himself and that the resemblance between the two works is the result of a "consistency of Ockeghem's style and melodic vocabulary."[23] Specifically, he points out that Ockeghem used the motive that underlies the openings of all the movements of the mass as well as the opening of chanson in a number of other pieces as well, including *Intemerata Dei mater* and *S'elle m'amera/Petite camusette.* If there is a relationship between *Missa Mi mi* and *Presque transi,* it is different from the one that exists between his other masses and their models in at least two important respects. *Missa Mi mi* does not use a cantus firmus in any of its movements, and none of the sources that transmit this mass links it with *Presque transi* in any way. As Christopher Reynolds has shown, however, numerous masses from the fifteenth century quote from chansons other than the one being used as a cantus firmus or when there is no secular cantus firmus in the first place,[24] so there is no obstacle to viewing *Missa Mi mi* as quoting from *Presque transi*.[25]

Before we examine examples from *Presque transi* and *Missa Mi mi*, I would like to reiterate a warning that Howard Brown made in a seminal essay on borrowing.

> Before asserting positively that such allusions were a part of the composer's conscious intent, we should attempt to demonstrate that the relationship between the model and the imitation is significant, that is, that a prominent theme in the model appears in a conspicuous place in the imitation, that the thematic material is individual enough for it to be recognized in a new context, . . . or that some other reason exists to support the claim of a significant relationship. Otherwise, given the amount of neutral melodic material common to all chansons in this period, it would not be difficult to demonstrate that every composition from the late fifteenth century is derived from every other.[26]

There are numerous putative quotations of *Presque transi* in *Missa Mi mi* that I do not believe would meet any of Brown's criteria. The examples in Figure 2.12 are indicative of these quotations: they are very short; they quote only a single voice; the rhythm has been significantly altered, often turning what was an important pitch in the chanson into a decorative note in the mass and vice versa; and the melodic material is, in my opinion, far too neutral to claim any relationship between the chanson and the mass. I

Fig. 2.12 (a) Ockeghem, *Presque transi,* superius, mm. 25–26; (b) Ockeghem, *Missa Mi mi,* Gloria, superius, mm. 119–22; (c) Ockeghem, *Presque transi,* superius, mm. 49–50; (d) Ockeghem, *Missa Mi mi,* Gloria, superius, mm. 145–46.

would make the same claim even if we were dealing with a *Missa Presque transi* that used a cantus firmus.

In addition to these probably fortuitous similarities, there are a number of quotations—both monophonic and polyphonic—that appear to be intentional. Altogether, there are only three polyphonic quotations of *Presque transi* in *Missa Mi mi,* one each in the Kyrie, Gloria, and Sanctus. They are brief, and literal or lightly ornamented. The most extensive quotations appear in the opening Kyrie (see Figure 2.13). In this passage there are four separate quotations. The opening three measures of the bassus quote the opening measures of the tenor of *Presque transi;* the quotation is very close to the original, with only a single passing note omitted. The superius of the mass quotes mm. 2–6 of the superius of the chanson; this quotation overlaps with the quotation in the bassus and in mm. 2–3 forms a polyphonic quotation in which the pitches and rhythms are the same, and the vertical alignment between the voices is maintained. The second half of the superius quotation makes a few small changes from the original by adding two pitches and slightly altering the rhythm; nevertheless, the relationship between the mass and chanson is still very clear here. The other two quotations are monophonic and shorter, but exact: m. 6 in the bassus quotes m. 12 of the tenor of *Presque transi,* and mm. 7–8 of the bassus quotes m. 16–17 of the tenor of the chanson. Measure 6 is not a polyphonic quotation but rather the overlapping of two monophonic quotations of material from different parts of the chanson. All of the quotations here, with their closeness to their original voices and their brevity, are similar to the way Ockeghem normally quotes from himself.

Other quotations of *Presque transi* in *Missa Mi mi* fall between those in Figures 2.12 and 2.13 in their use of borrowed material. For example, there are numerous reputed monophonic quotations of the superius from the opening of the second half *Presque transi* (mm. 32–37), but most of these are short and paraphrased to some degree, making their relationship to *Presque transi* tenuous at best.[27] Two such passages are given in Figure 2.14. In my opinion, Figure 2.14b is not based on *Presque transi.* Although it has the same starting pitch, the same high note, and a somewhat similar contour, the resemblance ends there. Many of the notes that gave the passage in *Presque transi* its distinctiveness have been removed, and the remaining melodic material is again too neutral to claim any relationship between the chanson and the mass. Furthermore, this passage does not exhibit the same approach to paraphrase that we have seen elsewhere in Ockeghem's music: Ockeghem expands melodic material when paraphrasing, while this example truncates it (compare this example with Figures 2.7 and 2.9). In contrast to Figure 2.14b, Figure 2.14c is very close to the original, especially after the second measure, and the few changes that occur are consistent with Ockeghem's manner of manipulating borrowed material. This quotation meets all of

(a)

Fig. 2.13 (a) Ockeghem, *Presque transi,* mm. 1–6; (b) Ockeghem, *Missa Mi mi,* Kyrie, mm. 1–8.

Fig. 2.14 (a) Ockeghem, *Presque transi,* mm. 32–37; (b) Ockeghem, *Missa Mi mi,* Credo, tenor, mm. 1–4; (c) Ockeghem, *Missa Mi mi,* Agnus, superius, mm. 1–6.

Brown's criteria: it comes from a prominent place in the chanson, it is conspicuously placed in the mass, and the quotation is clearly recognizable.

It is clear that Ockeghem did make some allusions to *Presque transi* in *Missa Mi mi,* but the number of these allusions is not as great as Miyazaki and others have suggested. The somewhat obscure relationship between these two works demonstrates the need for more in-depth studies of personal style in general and borrowing procedures in particular. Ockeghem seems to have had one approach to borrowing when he was using someone else's material but a rather different approach when he was using his own: when borrowing from others he had to interpret the material to suit his needs, but when borrowing from himself he did not. Knowing this helps us understand the relationship between *Missa Mi mi* and *Presque transi* much better.

Ockeghem's *Missa L'homme armé* and *L'aultre d'antan* and Robert Morton's *Il sera pour vous/L'homme armé*

The last of Ockeghem's masses that is indebted to a polyphonic model is *Missa L'homme armé.* Leeman Perkins suggests that it is not based on the monophonic tune but rather jointly on Ockeghem's own rondeau cinquain *L'aultre d'antan* and Robert Morton's *Il sera pour vous/L'homme armé.*[28] Perkins posits an extensive network of relationships among these works (and Busnoys's *Missa L'homme armé* as well) that includes not only melodic correspondences but also comparable notational peculiarities and similar modal and mensural treatment. In regard to the melodic correspondence, Perkins claims that Ockeghem's *Missa L'homme armé* quotes from three sections of *Il sera pour vous/L'homme armé:* the incipit of the cantus part,

Fig. 2.15 (a) Ockeghem, *L'aultre d'antan,* tenor, mm. 2–5; (b) Ockeghem, *Missa L'homme armé,* Kyrie, superius, mm. 1–2; (c) Ockeghem, *Missa L'homme armé,* Gloria, bassus, mm. 125–26; (d) Ockeghem, *Missa L'homme armé,* Agnus, altus, mm. 47–49.

the cadential phrase associated with the words "le doubté Turcq" (mm. 4–5 of Morton's chanson), and the fanfare figure in close imitation that occurs frequently in the Morton (mm. 3–4, 5–6, 7–8, and 14–15).[29] Perkins also cites numerous quotations of the opening of *L'aultre d'antan.* Many of the melodic correspondences, however, do not meet any criteria for showing that they were consciously done, suggesting that the melodic correspondence between *Missa L'homme armé, L'aultre d'antan,* and *Il sera pour vous/L'homme armé* may not be as strong as formerly thought.

Figure 2.15 compares the opening of *L'aultre d'antan* with its supposed use in *Missa L'homme armé.* The melodic material—a stepwise ascent of a sixth—is moderately distinctive, though Ockeghem used a similar phrase in his chansons *Fors seulement l'actente* (mm. 44–45) and *Ma bouche rit* (mm. 37–39).[30] In Figures 2.15b and 2.15c, however, he used different portions of that ascent (a fourth and a fifth, respectively), which are far more common and once again not individual enough to substantiate the claim of derivation. Furthermore, the supposed quotations are very short, ranging only from four to eight notes (one to three measures); each quotation is rhythmically altered; and the second of the three quotations (Figure 2.15c), appearing in the middle of the final section of the Gloria, is not prominently positioned. In Figure 2.15d the quotation encompasses the entire rising sixth, but there are significant rhythmic alterations, and on the whole this example is as similar to the passages in *Fors seulement l'actente* and *Ma bouche rit* as to

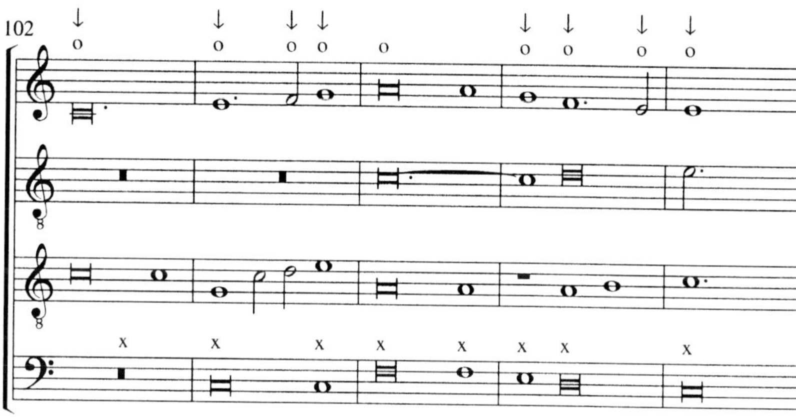

Fig. 2.16 Ockeghem, *Missa L'homme armé*, Credo, mm. 102–6.

L'aultre d'antan. The most we can say about these examples is that they too represent a consistency in Ockeghem's melodic vocabulary.

A similar case can be made against most of the putative quotations of *Il sera pour vous/L'homme armé,* and for the same reasons. The cadential phrase associated with the words "le doubté Turcq" is exactly that: a very common melodic phrase associated with the superius line in many cadences. Ockeghem uses it—or a slightly varied version of it—in numerous songs, as do many of his contemporaries, including Dufay and Busnoys.[31] The same reservations can be raised concerning most of the supposed quotations of the incipit of the superius. There is, however, one quotation that is very clearly derived from the Morton chanson (see Figure 2.16).[32] Indeed, this is the only polyphonic quotation in *Missa L'homme armé.* In this passage Ockeghem quotes the first three measures of both the superius and tenor of Morton's chanson. Although it is clear that the quotation in the bassus ultimately derives from the monophonic tune, it seems highly unlikely that the quotation in the superius is merely fortuitous. With only two minor changes (one rhythmic alteration and one repeated note), it is very close to Morton's original, and the vertical alignment between the two voices has been maintained.

To summarize the relationship between *Missa L'homme armé* and its possible models, the mass does not appear to have any direct quotations of *L'aultre d'antan* and only a single quotation from Morton's *Il sera pour vous/L'homme armé.* In regard to the fanfare figure in close imitation, the passage cited by Perkins (Credo, mm. 78–80)[33] represents not an actual quotation of either chanson but rather the use of a similar technique, and any resemblance between *Missa L'homme armé* and *L'aultre d'antan* therefore is simply another example of a consistency in Ockeghem's melodic vocabulary. Furthermore, it has yet to be proven which of the two is the earlier piece. Fitch has argued persuasively for an early date for the mass, and so it is possible that it predates the chanson.[34]

It is reasonable to conclude that original source for the cantus firmus of Ockeghem's *Missa L'homme arme* is neither his polyphonic chanson *L'aultre d'antan* nor Morton's *Il sera pour vous/L'homme armé,* but rather the monophonic tune. If the model had been either of the polyphonic chansons, we would expect Ockeghem to have used more polyphonic quotations than are present in *Missa L'homme armé.* There are no polyphonic quotations of *L'aultre d'antan,* and the one, brief quotation of *Il sera pour vous/L'homme armé* is simply an allusion to that chanson. Ockeghem does the same thing in *Missa Mi mi,* where he alludes to a chanson that does not serve as the model for the mass.

* * *

No two Renaissance composers deal with borrowing—either monophonic or polyphonic—in precisely the same way. Some aspects of borrowing—such as the length and frequency of quotations—frequently change, not only from composer to composer but also within a single composer's oeuvre. For example, some of Ockeghem's polyphonic borrowings are short quotations of only a measure or two, whereas others are longer and encompass an entire phrase of the model. Furthermore, when Ockeghem based a mass on a polyphonic model, he usually used at least one polyphonic quotation in each movement, often more. His masses not based on a polyphonic model per se but that make allusions to another work contain fewer polyphonic quotations.

Other aspects of borrowing may vary from composer to composer but remain relatively stable for individual composers. As noted earlier, during the second half of the fifteenth century polyphonic quotations were of three different kinds: literal (occasionally with the addition of a few ornamental notes), paraphrased, or mixed (a combination of both literal and paraphrased quotations together).[35] These three types of polyphonic quotation were used by dozens of composers, beginning with Dufay in his *Missa Se la face ay pale* and continuing through the end of the fifteenth century. Furthermore, composers did not use several methods during their careers, but rather employed only one of the methods in all of their masses with polyphonic quotations. For example, Johannes Martini always interpolated polyphonic quotations literally, as in his *Missa La martinella;* similarly, Isaac only used paraphrased quotations in his masses based on polyphonic models, and Josquin always combined literal and paraphrased quotations. Occasional exceptions are found, as when Isaac quotes the opening of Bosfrin's

Et trop penser literally in the Sanctus of his *Missa Et trop penser,* but these are infrequent and do not subvert the basic hypothesis.

Ockeghem, on the other hand, is the only composer who varied his approach to borrowed material. When he based a mass on his own work, he borrowed literally (*Missa Fors seulement, Missa Ma maistresse,* and *Missa Au travail suis*); however, when he based a mass on another composer's work, he paraphrased melodic material (*Missa De plus en plus*). In contrast, Ockeghem borrowed literally in allusions, whether the allusion is to one of his own works (*Missa Mi mi, Intemerata Dei mater*) or to someone else's work (*Missa L'homme armé*).

Two possible explanations for this intertextual dichotomy are immediately apparent. First, *Missa De plus en plus*—the one work that significantly paraphrases its model—is probably an early work; indeed, it may be his first mass based on a chanson.[36] As such, it may represent Ockeghem's first thoughts on using a polyphonic model that he significantly revised in his later masses. Second, he may have used paraphrase in *Missa De plus en plus* simply because its model was written by another composer. The hypothesis that Ockeghem paraphrased music when he based a mass on another composer's model but remained much closer to the original when he based a mass on his own works supports the attribution of *Au travail suis* to Ockeghem. This hypothesis is not contradicted by the use of only a literal quotation of Morton's *Il sera pour vous/L'homme armé* in Ockeghem's *Missa L'homme armé.* Masses based on polyphonic models have far more references to their models than masses that only allude to a model. This affords the composer the opportunity to use paraphrase while minimizing the risk that references will be unrecognizable. When a composer only alludes to a model, he does not have that luxury, and he may consciously make quotations literal in order to keep them recognizable. Because these hypotheses are not mutually exclusive, it may be that both of them obtain. Because we have no other mass written later in Ockeghem's career that is based on a polyphonic model by another composer, neither hypothesis can be proven conclusively. Regardless of which hypothesis is correct, however, Ockeghem's approach to polyphonic borrowing remains unique in the second half of the fifteenth century.

Finally, the study of borrowing has ramifications far beyond simply knowing the location and type of quotations for an individual mass. As shown above with the masses based on unproblematic models, Ockeghem used literal quotations when borrowing from himself but polyphonic quotations when borrowing from others, strongly suggesting that in his *Missa Au travail suis* he is borrowing from his own chanson. Similarly, the newly detailed knowledge of Ockeghem's borrowing profile presented in this essay helps solve the mystery of his model for *Missa L'homme armé:* only the use of a monophonic model would have generated the kind of borrowing that Ockeghem employs in this famous and controversial work. The close study of an individual's use of polyphonic borrowing is thus a new and powerful tool for solving some of the many puzzles of fifteenth-century music.

Notes

1. Murray Steib, "Imitation and Elaboration: The Use of Borrowed Material in Masses from the Late Fifteenth Century" (Ph.D. dissertation, University of Chicago, 1992); and Steib, "A Composer Looks at His Model: Polyphonic Borrowing in Masses from the Late Fifteenth Century," *TVNM* 46 (1996): 5–41.
2. The authenticity of some of these has been questioned; see Andrew Kirkman, *The Three-Voice Mass in the Later Fifteenth and Early Sixteenth Centuries: Style, Distribution, and Case Studies* (New York: Garland, 1995), 249–60; and Fabrice Fitch, *Johannes Ockeghem: Masses and Models,* Collection Ricercar 2 (Paris: Honoré Champion, 1997), 177–78.
3. Even as recent and important a book as Fitch, *Johannes Ockeghem,* does not discuss polyphonic borrowing in much detail.
4. Jeffrey Dean, "Okeghem's Valediction? The Meaning of *Intemerata Dei mater,*" in Philippe Vendrix, ed., *Johannes Ockeghem: Actes du XLe Colloque international d'études humanistes, Tours, 3–8 février 1997* ([Paris]: Klincksieck, 1998), 521–70.
5. Christopher A. Reynolds, *Papal Patronage and the Music of St. Peter's, 1380–1513* (Berkeley, Los Angeles, and London: University of California Press, 1995), 258–62.
6. Howard Mayer Brown, "Notes Towards a Definition of Personal Style: Conflicting Attributions and the Six-Part Motets of Josquin and Mouton," in Willem Elders with Frits de Haen, eds., *Proceedings of the International Josquin Symposium, Utrecht 1986* (Utrecht: Vereniging voor Nederlandse Muziekgeschiedenis, 1991), 185–207, at 189.
7. Some of the earliest uses of this term in musicological studies include Paula Marie Higgins, "Antoine Busnois and Musical Culture in Late Fifteenth-Century France and Burgundy" (Ph.D. dissertation, Princeton University, 1987); Rob C. Wegman, "Another 'Imitation' of Busnoys's *Missa L'Homme armé*—and Some Observations on *Imitatio* in Renaissance Music," *JRMA* 114 (1989): 189–202; Michael Allsen, "Style and Intertextuality in the Isorhythmic Motet, 1400–1440" (Ph.D. dissertation, University of Wisconsin, Madison, 1992); and David Crook, *Orlando di Lasso's Imitation Magnificats for Counter-Reformation Munich* (Princeton: Princeton University Press, 1994). Julia Kristeva, "Bakhtine, le mot, le dialogue et le roman," *Critique* 23 (1967): 438–65, was the first to use the term, according to Crook, *Orlando di Lasso's Imitation Magnificats,* 155.
8. Although *Fors seulement l'actente* is a rondeau cinquain, two of the lines of text consist of two musical phrases each, which accounts for the seven phrases of music listed in Figure 2.2. Voice designations follow those in Johannes Ockeghem, *Masses and Mass Sections,* ed. Jaap van Benthem (Utrecht: Koninklijke Vereniging voor Nederlandse Muziekgeschiedenis, 1994–), II/4: vii–ix, where the seven phrases are distributed as follows: A, mm. 1–10; B, 10–18; C, 18–30; D, 30–41; E, 42–47; F, 47–55; G, 55–70.
9. Unless otherwise indicated, measure numbers refer to Ockeghem, *Masses and Mass Sections.*
10. In this and all subsequent music examples, the symbols used have the following meanings:

    PT = passing tone
    x = a quotation from the tenor of the model
    o = a quotation from the superius of the model
    \+ = a quotation from the contratenor of the model
    ~ = one or more notes of the model have been omitted
    ↓ or ↑ = the vertical alignment between this voice and the voice with the cantus firmus has been maintained for this pitch (the arrow always points toward the cantus firmus)

11. Another example of this occurs in mm. 7–10 of the Gloria, where he combines a portion of phrase B of the tenor with the opening of phrase A of the discantus (in the bassus and superius, respectively).

12. Measure numbers for this mass refer to Johannes Ockeghem, *Collected Works,* 3 vols., 2nd ed. of vols. 1–2, ed. Dragan Plamenac and Richard Wexler, American Musicological Society Studies and Documents 1, 3, 7 (n.p.: American Musicological Society, 1959–92), 1:117–23.
13. Fitch, *Johannes Ockeghem,* 76 and passim.
14. In these cases, such repetitions are notated with an *x* or *o* in parentheses [(x) or (o)], and a dotted line connecting them to the original quoted note.
15. Four of the fifteen polyphonic citations quote the opening of the chanson.
16. Steib, "A Composer Looks at His Model," 5–41.
17. Leeman L. Perkins, "Ockeghem, Jean de," *New Grove II.*
18. David Fallows, "Johannes Ockeghem: The Changing Image, the Songs, and a New Source," *Early Music* 12 (1984): 218–30, at 223–25.
19. Ockeghem, *Collected Works,* 3:ciii.
20. Fitch, *Johannes Ockeghem,* 157–58.
21. Measure numbers for this mass refer to Ockeghem, *Collected Works,* 1:30–41.
22. Haruyo Miyazaki, "New Light on Ockeghem's *Missa Mi-mi,*" *Early Music* 13 (1985): 367–75; Wolfgang Thein, *Musikalischer Satz und Textdarbietung im Werk von Johannes Ockeghem* Würzburger musikhistorische Beiträge 13 (Tutzing: Hans Schneider, 1992), 115–20; Rebecca Stewart, "*. . . Ita desiderat anima mea ad te, Deus* (Ps. 42:1)*:* Johannes Ockeghem, a Most Medieval Musician," *TVNM* 47 (1997): 163–200, at 170–95; Fitch, *Johannes Ockeghem,* 160–71; van Benthem in Ockeghem, *Masses and Mass Sections,* III/2:xii–xiii.
23. Martin Picker, "Reflections on Ockeghem and *Mi-Mi,*" in Philippe Vendrix, ed., *Johannes Ockeghem: Actes du XL*[e] *Colloque international d'études humanistes, Tours, 3–8 février 1997* ([Paris]: Klincksieck, 1998), 415–32, at 423.
24. Christopher A. Reynolds, "The Counterpoint of Allusion in Fifteenth-Century Masses," *JAMS* 45 (1992): 228–60; and Reynolds, *Papal Patronage,* 249–97.
25. Fitch, *Johannes Ockeghem,* 159–61, talks about the differing opinions concerning the relationship between *Missa Mi mi* and *Presque transi* at greater length.
26. Howard Mayer Brown, "Emulation, Competition, and Homage: Imitation and Theories of Imitation in the Renaissance," *JAMS* 35 (1982): 1–48, at 14. Other writers have given similar criteria; see also Reynolds, "The Counterpoint of Allusion," 247; Leeman L. Perkins, "The L'Homme Armé Masses of Busnoys and Okeghem: A Comparison," *Journal of Musicology* 3 (1984): 363–96, at 381; and Leo Treitler, "Music Analysis in an Historical Context," *College Music Symposium* 6 (1966): 75–88, at 79–82.
27. Many of these are given in Miyazaki, "New Light on Ockeghem's *Missa 'Mi-mi,'*" 373.
28. Perkins, "The L'Homme Armé Masses," 377–81.
29. For an edition of this chanson, see Leeman L. Perkins and Howard Garey, eds., *The Mellon Chansonnier,* (New Haven and London: Yale University Press, 1979), 2:125.
30. The first of these examples is rhythmically different, but the second is similar. For an edition of *Ma bouche rit,* see Ockeghem, *Collected Works,* 3:73–74.
31. For example, see the following songs by Ockeghem: *Aultre Venus* (mm. 15–17), *Fors seullement contre* (36–38), *Prenez sur moi* (33–35), *Presque transi* (29–31), *Quant de vous seul* (34–36), and *Qu'es mi vida* (32–34). For editions of these works, see Ockeghem, *Collected Works,* 3: 59, 64–65, 80, 81–82, 83, and 84–85, respectively.
32. This is the only polyphonic quotation of either Morton's *Il sera pour vous/L'homme armé* or Ockeghem's *L'aultre d'antan* that I have found in this mass.
33. Perkins, "The L'Homme Armé Masses," 378.
34. Fitch, *Johannes Ockeghem,* 56–61. As always, when dealing with Ockeghem, it is nearly impossible to put specific dates to individual pieces. Currently, the best we can hope for is to date individual works relative to one another.
35. It is important to note that composers who use both literal and paraphrased quotations do so in each mass, rather than using literal quotations in some masses and paraphrased quotations in others; see Steib, "Imitation and Elaboration," chapter 4, and Steib, "A Composer Looks at His Model."
36. Fitch, *Johannes Ockeghem,* 77.

3
The Illusion of Allusion

JENNY HODGSON

The terminological difficulties repeatedly encountered in musicological writings on borrowing are unlikely to be resolved in this volume.[1] This is perhaps how the situation should remain, for the crux of the issue lies in the fact that identified examples of borrowing often vary dramatically in their relationship to the source material(s) from which they have been derived. As a result, networks of affiliations are difficult to establish. The hypothesis advanced in this paper will suggest yet another relationship, although, in a slight twist to the issue, the source materials will be identified only tentatively. Furthermore, it will remain an issue of conjecture as to whether the borrowings were the result of conscious thought or habitual and reiterative musical practices. The latter possibility, seeming more applicable to a study of musical style, is discussed here primarily in response to research in which certain compositional practices have been assumed to be demonstrative of the text-based borrowing procedure known as the "contrapuntal allusion."[2]

More specifically, though, the central processes with which this study shall be concerned are those that are likely to have been used during the performance of extemporized polyphony. It will be suggested that residua of these processes—perhaps even direct referrals—exist within the notated polyphony of the fifteenth century. This residue will be shown to constitute many of the fundamental contrapuntal structures identified by previous writers as "clichéd" or "allusory." From this basis, we will identify some of the techniques by which singers—as a collective body—coordinated consonances when singing discant. In this light, the passages will still be considered "borrowings." The main difference will be that they cannot be considered to be the product of any individual; the source of the borrowings is therefore understood as the community itself.

Owing to the nature of the phenomenon, of course, very little is known with certainty about these techniques. Although counterpoint treatises of the time tell us all we need to know about the singing of consonant intervals, they are effectively silent with regard to the methods by which singers coordinated their efforts into larger successions of perfections, and from there to larger phrases, and so on.[3] To confound matters even further, the nature of the relationship between these practices and the works that survive in notated form remains decidedly vague. Vague, that is, but not entirely inaccessible.

Clues are rare, but they do exist. Scribal alterations to chansons indicate that notated works were not necessarily considered "fixed" once they were committed to paper. Superius lines are particularly susceptible to seemingly minor alterations such as changes in cadential approaches and modest rhythmic details. Howard Mayer Brown and others have suggested that these may reflect a certain degree of flexibility in performance, particularly with regard to written-out embellishments.[4] This is surely borne out by the embellishment manuals of the mid-sixteenth century, the basic purpose of which is to provide typical melodic progressions and methods of embellishing (i.e., discanting) over them.[5] Lack of multiple sources for masses, however, has hindered the extensive study of embellishment procedures in this genre. Nonetheless, the scribal variants we do have typically pertain to the same features as those found in chansons: superius lines and cadences. From this and what follows it may be presumed that a certain degree of flexibility was common to both mass and chanson performance—if not always in actual performance, then at least by way of alterations introduced by scribes.

An anonymous *Da pacem* antiphon in VatSP B80 gives some intimation of what this flexibility may have entailed, in this case at the Basilica of St. Peter's (Rome) during the mid-1470s.[6] The work survives in two versions: a primary rendering in which the antiphon is presented in a relatively simple form, and a later redaction in which various elaborations have been added to the superius line. This line is not newly composed, as is often the case in chanson redactions, nor is it simply an ornamented variant of the original. Instead, the degree of embellishment added to the line varies between the two extremes, that is, simple elaboration of the original and full recomposition. This and the fact that the superius is supported by a relatively simple tenor-bassus framework would suggest that the embellishments are likely to be reflective of improvisatory practices. As the variants affect only a single vocal line, one could imagine that embellishments of this type could have easily been added during a performance marked only by very casual prior planning. It is therefore possible that the scribe was "modernizing" the original manuscript copy of the work according to current improvisatory styles at the institution.[7] In this light, the "variant" reading thus belongs as much to the notated realm as to the improvisatory one in which it was engendered.[8]

Further evidence for a strong relationship between the use of improvisatory gestures in sacred music and those that were notated is provided by fragments of didactic exercises comparable to those mentioned earlier. The fragments contain chant, rather than chanson, tenors that have been supplied with embellishments similar to those in the later diminution manuals.[9] These are important because the embellishments in these fragments are clearly more appropriate for sacred genres than song performance.[10] In this respect it is perhaps interesting to note that polyphonic settings of the mass are not found in the printed embellishment manuals: examples therein are limited to chansons, madrigals, and motets.[11] Yet there can be little doubt that the singing of discant during the Mass Ordinary was a widespread occurence. Paula Higgins, for instance, has shown that the statutes of the Bourges Palace prescribe that discant be performed daily "at every mass of whatsoever solemnity."[12] Rob Wegman's recent study has underlined such observations by showing that discant was an almost ubiquitous practice in the churches and courts of the fifteenth century.[13] What has been little commented upon, however, is the fact that much of our knowledge of improvisatory practices suggests that it was highly formulaic in nature. More detailed evidence for this conclusion shall be supplied below, but for the moment let us remain with the issue of embellishment exercises and scribal variants.

Central to both phenomena are standardized melodic patterns. The embellishment exercises consist of little more. Scribal variants, on the other hand, firmly underline the fact that such patterns were highly institutionalized: within the work of a single scribe (or group of scribes, if working at the same institution), specific preferences are typically found with regard to cadential approaches, the splitting of sustained notes, and so on.[14] But while scribal variants point to little more than the alteration of these limited features, the embellishment exercises have no such confines. The techniques they utilize are clearly intended as "starters" for apprentices new to the art of singing *cantare super librum*. In order to develop fluency when discanting upon these short segments, the singer may have performed exercises akin to those suggested by Vicente Lusitano in 1553. He recommended that when improvising over a familiar skeletal counterpoint the singer should "[s]ing a melodic pattern once or twice, and immediately afterwards a run *(tirata)*, or a large skip (*passo largo*) up or down, as you see fit."[15] Once proficient in discanting in this way, the skills of the singer would have been applied to entire chants, rather than chant segments. With regard to discant-based gestures, we may conclude that the formulas applied to discrete melodic units could therefore be expected to be similarly utilized in extended chant improvisations.

In this sense the didactic exercises provide our second clue about the techniques by which singers coordinated their efforts when singing *super librum.* By constructing the discanting (i.e., superius) line from a series of preconceived melodic figures, the singer would always have had solutions to hand for dealing with the tenor progressions upon which the improvisation was based. To borrow an analogy from chant studies, these stereotyped forms can be thought of as "signposts" by which performances were organized.[16] For polyphony, the coordination of one or more contrapuntal voices would surely be supported by this framework, in that the formulas utilized by the superius would also be familiar to additional singers and thus serve to direct them with regard to the harmonic progressions formed between that voice and the tenor. For nonmensural polyphony, the art of formulas has been tentatively suggested by several writers, but few have attempted to identify specific formulas for the reason that the sources that contain this music require extensive examination to enable systematic comparison between the many localized repertories: quantity, liturgical function, dating, provenance, and contents all require categorization.[17]

It takes only a small but significant step to connect these standardized patterns with those that Reynolds and others have identified as melodic or "contrapuntal allusions." Quite simply, these allusions denote a process by which composers of the mid- to late fifteenth century quoted or paraphrased short melodic phrases from each other with the intent of establishing a musico-textual allusion between their work and its model. The significance of some such quotations are unquestionable: many chansons, for instance, can be shown to share both initial text and music.[18] Such correspondences are more difficult to establish in the case of mass settings. This is primarily owing to the standardized text and the sheer preponderance of apparently standardized melodic gestures. Rather than being restricted to their structural foundations—perhaps a shared head motive or similar tenor—these oft-repeated gestures can be found to accompany any portion of the mass text and furthermore utilize anything from one to all four contrapuntal parts of the "model." As we shall observe in this study, Reynolds effectively diluted the value of his own observations by extending allusion theory from chansons to polyphonic masses.

Although some of Reynolds's examples feature a handful of shared melodic or contrapuntal phrases, the majority of the allusions consist only of a *single* (often monophonic) phrase. Despite being only one of several hundred found within any typical mass cycle, this phrase may nonetheless be considered consequential if it corresponds to one of a similar (but rarely the same) melodic outline in a chanson or mass of the period. In order to safeguard against accusations of insularity, Reynolds categorizes melodic correspondences in one of three ways. They are deemed allusory if the texts are theologically concordant, contrapuntal citations if they occur within the context of "substantive motific ties," and mere clichés otherwise.[19] Yet while the obviously arbitrary nature of pinpointing this or that melodic type was not lost on Reynolds, it barely hides the fact that few of his allusions extend beyond a perfection or two. More seriously, these are identified at the expense of other recurrent motives that are simply discarded if found lacking in textual significance or sheer weight in numbers.

In this respect more than any other, Reynolds's analytical framework seems woefully inadequate. Yet his study is nonetheless useful in that it identifies complexes of works that rely heavily upon successions of standardized melodic and contrapuntal gestures. Interestingly, while these gestures are present in some works, they are clearly conspicuously absent in others. Furthermore, the allusions identified by Reynolds seem to concur well with the properties observed in embellishment exercises and works affected by scribal alterations. Standardized albeit institutionalized gestures are crucial in each case. But the similarities do not end there. Like the scribal variants, the allusions are typically found in the superius lines both of chansons and masses. Their length, meanwhile, is perhaps more reflective of the embellishment formulas: although typically short (often no more than two perfections), they can occasionally be extended to encompass entire phrases.

It would be a mistake, though, to directly equate the recurrent melodic and contrapuntal gestures found in such polyphony with that of an ostensibly "lost" improvisatory tradition. Although the gestures were indeed highly stylized in some respects, the didactic exercises indicate that one embellishment could have been as easily used as another when filling in a progression prescribed by the chant tenor. It is therefore the basic structural qualities of the gestures that are all-important. As a result, their functional significance must be proved before they can be presumed to be reflective of improvisatory practices. Without functionality the contrapuntal frameworks become simply elements of syntax. Once established, however, the precise details of the polyphony surely become significant. As with scribal idiosyncrasies, greater concordances between melodic embellishments are likely to indicate correspondences of a geographic or chronological nature.

But if recurrent gestures can be shown to possess functionality reflective of the coordinative procedures likely to have been used in extemporization, such an analysis invariably presumes something about the essence of the work in which they are to be found. That is to say, any judgment will by necessity place that work at some point on the continuum between improvisation and composition. As Margaret Bent has argued, any audible distinction between the two is likely to have been minimal if those singing *cantare super librum* successfully followed the rules of counterpoint.[20] As a result, there might be little stylistic difference between the coordination of unwritten counterpoint in performance and the coordination of simultaneities in the mind of the individual *compositor* prior to their notation. As Rob Wegman concluded, the distinction between the presence or absence of writing in the compositional process is likely to have been purely empirical.[21] An important conceptual difference, however, is that *cantare super librum* clearly required much prior planning to ensure that consonances were successfully resolved. On the other hand, *res facta* was of necessity "correct" because it was the result of the application of the rules of counterpoint. In this respect, we might therefore equate the contrapuntal and melodic formulas discussed in allusion studies with those techniques that singers used in *cantare super librum* to coordinate their collective efforts, and thus "avoid similarity between . . . the choice and ordering of concords."[22] In consequence, the utilization of such procedures within polyphony (be it notated or otherwise) should surely be considered reflective less of improvisation or composition than of a continuum between institutionalism and individualism—in other words, between works written in conformity with localized or institutionalized extemporization practices and those in which the preferences of the individual (i.e., the composer or scribe) may have played a more dominant role.

This distinction is further supported by the issue mentioned earlier concerning the precise methods by which singers performed discant. The reason we know so little about these is because counterpoint treatises of the period are unanimously silent concerning the specific methods by which consonances should be organized. It would seem that their reticence most likely reflects exigencies pertaining to the codification of the practice rather than any vagaries of the theoretical tradition. As Richard Crocker has noted, the treatises of the fifteenth century reveal "no basic novelty when compared to earlier sources" in that they exhibit "no change in the basic principles" of discant.[23] As the practice appears to have been a highly structured one with regard to the degree of prior planning required for successful performance, this leaves only one possibility: that the art of coordinating simultaneities when singing *super librum* was highly localized and therefore, unlike the counterpoint that it framed, could not be codified. The highly generalized instructions of Lusitano and others undoubtedly reflect this phenomenon. If discant practices were indeed highly institutionalized, then we must furthermore conclude that works that rely heavily upon stylized gestures as a basis of organizing polyphony are likely to reflect institutional norms to a far higher degree than those in which such formulas are absent.

The evaluation of fifteenth-century polyphony in terms of the institutions for which it was created has proved particularly advantageous when it comes to understanding the structural peculiarities of sacred polyphony. Most research has centered on the idiosyncratic liturgical practices for which these works were composed (for more information, see below). In terms of the individual/institution dichotomy, however, one specific set of works has proven particularly difficult for scholars to fathom on both counts: the set of six *L'homme armé* masses preserved in the manuscript NapBN 40 of the Biblioteca Nazionale, Naples.[24] Despite many suggestions, no plausible explanation has been provided for their genesis, and no individual has been convincingly credited with their authorship. Are they the product of an individual (Taruskin, Giller) or of many hands (Cohen, Haagh)? Are they progressive (Cohen) or simply archaic and essentially hackneyed (Giller, Reynolds)?[25]

Part of the difficulty in understanding the Naples masses surely lies in the boldness of their conception: six cantus firmus masses connected by a single ground plan. Furthermore, unlike the only large-scale precedents for such a project (the eight Introit settings by Johannes Brassart, the Mass Proper cycles in TrentC 88, and the cycles of hymns and Kyrie settings by Dufay), the individual masses of the complex paradoxically display little sense of belonging to a cycle. Aside from the common cantus firmus, there seem to be few other explicit correspondences (such as mensural idiosyncrasies, shared head motives, etc.) that cannot otherwise be explained in terms of stylistic cliches.

The situation is no better with attempts to define the complex in terms of authorship. When Leo Treitler asked what critics are "to do in the face of works that appear to recompose old stuff and bear the mark of many, or of no, hands?" he might well have been referring to the Naples mass complex.[26] The difficulties that scholars have experienced in this respect are surely exemplified by Judith Cohen's early analysis: cantus firmus treatment is like the early masses of Dufay, sequential progressions like those of Obrecht and Busnoys (Masses II and III), cantus firmus canon scaffolding like that of Faugues (Mass VI), use of cantus firmus incipit in opening mottos like that of Busnoys, and so on.[27] By contrast, Don Giller has made an intriguing argument for Caron's authorship of the set, an evaluation made on the basis of a more generalized stylistic analysis.[28] This particular relationship will be discussed to a limited extent in the current study, although authorship per se shall not be an objective.

The hackneyed melodic gestures and commonplace contrapuntal frameworks that dominate the Naples set would seem to be poor substitutes for the telltale notational practices and compositional idiosyncrasies that typically prove so helpful in attribution studies. Yet if the stylized features common to the masses are indeed reflective of extemporization practices, then it is possible that they could also provide the set with the institutional—if not authorial—fingerprint we have been looking for. The studies of Cohen, Giller, and Reynolds all point to the presence of such features, although each with different objectives in mind.[29] Care must be exercised in this respect, though, as the identification of melodic or contrapuntal commonplaces is fraught with conceptual difficulties. As in studies of medieval chant, the identification of meaningful correspondences within allusion studies has been hampered by the lack of an empirical methodology. But as Peter Jeffery has noted for chant repertories, this problem should be understood as a consequence of the nature of the phenomenon rather than any deficiency in modern scholarly procedures. The recall of formulas is unlikely to have been a wholly conscious act and therefore should be understood in terms of the individual and/or stylistic idiom in which the singer/composer was operating.[30] In this respect it is useful to bear in mind Rob Wegman's assertion that any interpretation of the nature of compositional process "amounts to the question of what, historically, composers thought and did before arriving at the notated work . . . [and] what, historically, singers thought and did when turning the notated work into the musical event."[31]

Identifying the functional qualities of a discant-based gesture tells something about these processes, the circumstances in which they were likely to have been utilized in performances of extemporized polyphony, and so on. So although it would be quite possible to identify innumerable examples of "characteristic" figures and their derivative forms, the inherent value in such an exercise—while perhaps beneficial in a study of larger scope—is surely minimal. Without functionality, the gestures become elements of syntax: individualistic features that can only denote the disassociation of a compositional style from its institutional, improvisatory origins. In this respect the relatively few gestures identified in this study are therefore discussed primarily in terms of their functionality rather than the particular qualities of their constitutive melodic figures. Although the length of the coordinating frameworks varies considerably, as does the extent to which melodic features are shared between them, the gestures chosen are nonetheless considered to be representative of those found throughout the Naples complex and certain related masses. It is perhaps most useful to think of these features as occupying a continuum whose extremes are occupied by repetition and recomposition.

The first example, shown in Figure 3.1, typifies those frameworks whose treatment occupies the midpoint on this continuum.[32] The melodic qualities of the two passages are not repeated verbatim, although they share enough features to make their common origins easily discernible. The passages are among five identified by Reynolds as having "associations" with Agnus II of Ockeghem's *Missa* (à 3).[33] According to Reynolds's allusion-based

Fig. 3.1 (a) Caron, *Missa L'homme armé,* Agnus Dei, mm. 43–53; (b) anonymous, *Missa L' homme armé* (Mass VI), Benedictus, mm. 1–15; transcription after *Six Anonymous L'homme armé Masses in Naples, Biblioteca Nazionale, MS VI E 40,* ed. Judith Cohen, CMM 85 (Neuhausen-Stuttgart: Hänssler/AIM, 1981), 279–80; used with permission.

interpretation, the measures from the opening of Caron's Agnus II are understood to have "provided the model" for the Benedictus from the sixth Naples mass. As this is the only borrowing that the masses seem to share, one might wonder why the melodic features of the contratenor are so disparate. If a quotation really was intended, it could hardly be less memorable or distinguished.[34] As Reynolds himself demonstrates, the features of the chantlike superius against which the contratenor discants are hardly uncommon. Don Giller, for instance, uses a phrase of the same melodic intervals to demonstrate the "thematic unity" of the Naples set.[35] It might be more convincing, therefore, to presume that both composers utilized a stock melodic counterpoint at this juncture. This would explain why the two passages share few melodic gestures (certainly none that would be audible) yet nonetheless maintain a common structural basis around which the counterpoint is coordinated. The precise details of the counterpoint differ in the way that the variant embellishments do in chant-based didactic exercises or diminution manuals. That both instances of the gesture (as well as those identified by Giller) occur at important structural divisions in the polyphony surely tells us something about the circumstances in which it was utilized in

Fig. 3.2 Anonymous, *Missa L' homme armé* (Mass II), Et in terra, mm. 45–46; transcription after *Six Anonymous L' homme armé Masses in Naples, Biblioteca Nazionale, MS VI E 40,* ed. Judith Cohen, CMM 85 (Neuhausen-Stuttgart: Hänssler/AIM, 1981), 56; used with permission.

extemporized polyphony. One possibility is that it could have been employed in performance to mark notable textual moments that were of particular liturgical significance. On the other hand, the occurance of the gesture at these points may simply reflect the fact that stereotyped forms—at least in chant and poetic narratives—typically occur at the beginnings and ends of constructions.[36]

With regard to the basic construction of the gesture, the fact that the contratenor discants against the superius surely gives us some clues as to how a commonly encountered cantus firmus progression might have engendered a stereotyped response. The long, chantlike notes in the superius of both passages are not explicitly derived from the *L'homme armé* melody and so must be understood to represent an "imitation" of an established, chant-based discant progression. Despite being neither a "real" chant segment nor presented by the "correct" voice part (see below), the superius is nonetheless given the status of the chant-bearing voice, in that the contratenor has coordinated its response to it as if it were a chant tenor. This may, of course, suggest that the chant-bearing voice might not always have been the tenor in discant-based improvisations. But this seems unlikely, as the tenor, in contrast to the other voice parts, appears to have played a singularly important role in the performance of extemporized polyphony.[37] The notion that this could not be accomplished without a tenorist suggests that this voice somehow led singers in coordinating improvisatory frameworks. In this light, an alternative explanation for the features found in Figure 3.1 might be that composers writing notated polyphony felt freer to move away from the improvisatory functions traditionally associated with specific voice parts.

Fig. 3.3 Caron, *Missa L' homme armé*, Patrem, mm. 48–50.

Coordinating structures (in two, three, and four parts) that share functional qualities similar to those given in Figure 3.1 are found throughout the Naples complex. All that typically differs is the degree to which the individual frameworks share specific melodic gestures. Correspondences between structures that share melodic outlines, of course, are most easily recognizable and have formed the majority of the stylistic comparisons made between the masses of Caron and the Naples set. Figure 3.2 shows a passage from the second Naples mass that can be considered a more extreme example among those found throughout the complex. Although this gesture is found in dozens of permutations, all variants share a relatively close adherence to the basic imitative motive found in the bassus of this example. All that typically differs are the precise points at which the arpeggios begin and end. The framework seems to be hyperpliable in this respect, in that innumerable variants are easily produced. It is only rarely that statements of the gesture fully correspond.

A case in point is the following passage from Caron's *Missa L'homme armé* (see Figure 3.3). This passage conforms almost exactly to the gesture cited in Figure 3.2. The exact function of the gesture and probable grounds for its sundry appearances are explained by the fact that its presentations (at least within the Naples complex and Caron's masses) are habitually associated with a specific musical act. They typically mark the entrance or withdrawal of the cantus firmus. More than just a commonplace, then, the precise function of the gesture may have been to provide a point of stasis at pivotal junctures when singing *super librum*. The interesting feature of the gesture, of course, is that it is not based on the cantus firmus in the way that most other frameworks are: the sustained cantus firmus note that typically accompanies it serves only to denote its incoming or outgoing presence. So it is perhaps significant that the gesture is rarely developed or embellished (as are many of the other discant-based frameworks found within the complex). Such

Fig. 3.4 Caron, *Missa Jesus autem transiens,* Qui tollis, mm. 88–95.

development never occurs within the Naples masses, although it is found once in Caron's oeuvre (see Figure 3.4).

As mentioned earlier, both Reynolds and Giller noted that particular melodic phrases occasionally appeared to be associated with similar structural or textual features: a motive might be consistently utilized as a head motive, or perhaps associated with a particular phrase of the mass text. The specific circumstances in which Figure 3.3 is typically employed further supports the conclusion that such correpondences owe less to allusions or thematic unity than to their coordinating functions in extemporized polyphony. Many of the correspondences Giller notes between the head motives of the Naples set and those in Caron's masses might be interpreted from this perspective. Just as Figures 3.3 and 3.4 herald the entrance/exit of the cantus firmus, so might these melodic gestures have been used to demarcate important textual divisions when extemporizing. Giller, for instance, identified a phrase ("Phrase 2") that began mass sections in all six Naples settings, four masses by Caron, and one apiece by Faugues and Dufay (see Figure 3.5).[38] Reynolds likewise identified many similar key phrases and their often abundant appearances in mass settings from ca. 1460 to 1480.[39]

Both studies naturally raise the question of whether melodic gestures might furthermore have become associated with particular textual phrases,

Fig. 3.5 Ascending melodic figure.

rather than nonspecific textual divisions. Connections between text phrases and melodic formulas are certainly not unknown in other genres. Some stereotyped textual phrases in Byzantine chant traditions seem to have been associated with specific melodic phrases,[40] while tentative suggestions have been made for similar practices in the Gregorian tradition.[41] As the embellishment manuals demonstrate, however, discant was by no means limited to genres in which texts were as rigidly fixed as those within liturgical practices. We can conclude from this that the frameworks utilized when discanting in each genre were in fact shared—at least within each localized tradition—and so it is unlikely that specific musico-textual connections could have developed between frameworks and the texts they declaimed. With regard to allusion theory, of course, this may provide an alternative explanation for the fact that contrapuntal frameworks are occasionally shared between chansons and masses. In terms of the works under consideration, this is further supported by the fact that the reuse of stereotyped phrases seems to correlate more with specific structural features rather than textual ones. Although Giller's Phrase 2 accompanies the text "Pleni sunt" in all six Naples masses, for instance, it is unlikely that this denotes a musico-textual connection for the reason that the phrase also accompanies a wide range of alternative textual divisions within the complex. Such practices are common both to Caron's masses and the Naples set.

Despite their many correspondences, the two mass complexes differ in one important respect. In Caron's masses the process of extending standardized contrapuntal frameworks is occasionally taken a stage further than in the Naples complex. This is particularly significant for those passages in which tenor statements are subjected to various degrees of rhythmic manipulation. Despite inconsistencies in the delivery of the cantus firmus, the contrapuntal frameworks are rigorously maintained. *Missa Clemens et benigna* is perhaps most notable in this respect. Despite a relatively free cantus firmus, the entire mass seems to consist of little more than a succession of discant-based frameworks. As in the Naples masses, the degree to which these coordinating structures are elaborated varies greatly. In some cases basic outlines are barely ornamented, in which case movement consists predominantly of breves. In others, these outlines can be transformed into filigree-style passages in which all voices move in minims and semiminims. The passages given in Figure 3.6 are typical.

The significance of the second passage in the example lies in the fact that the framework itself is utilized in a musical situation that bears little resemblance to any that is likely to have been encountered when singing *super librum*. The abrupt diminution of the tenor pitches is not exact and so introduces an element of inconsistency into the presentation; singers are unlikely to have been able to deal with this when extemporizing. This and

Fig. 3.6 Caron, *Missa Clemens et benigna* (a) Osanna I, mm. 55–62; (b) Confiteor, mm. 236–38.

similar passages in *Missa Clemens* therefore demonstrate that discant-based gestures could be carried over into portions of "composed" works whose structural foundations might bear little stylistic resemblance to those from which they originated. In this respect, we might conclude that these features are disassociated enough from actual discant practices to be considered "individual"—as opposed to institutional—features of Caron's compositional style. There is some evidence to suggest that this detachment may even have extended to his chansons. Certain passages from Caron's *Pour regard doeul* and his *Missa Accueilly m'a la belle* share a relationship similar to that observed in *Missa Clemens* (see Figure 3.7).

On the basis of this and other motivic relations between *Pour regard doeul* and the *Missa Accueilly m'a la belle*, Reynolds suggests that the mass is "dependent" upon the chanson in two ways: motivically and textually.[42] But as Reynolds himself comments, the chanson seems to share melodic and contrapuntal material with a variety of contemporary mass settings: the anonymous mass (VatSP B80, fols. 122–29) that presumably also contains

Fig. 3.7 Caron (a) *Pour regard doeul*, mm. 1–6; (b) *Missa Accueilly m'a la belle*, Benedictus, mm. 139–48.

material from the sixth Naples mass (see Figure 3.1, above), Agnus II of the third mass from the complex, and the Patrem from Ockeghem's *Missa Ecce ancilla domini.*[43] It might be more feasible to conclude that the chanson simply owes more to discant-based frameworks and embellishment patterns than other works of its genre. Caron's pronounced reliance upon such techniques in the masses *Accueilly m'a la belle* and *Clemens et benigna*, in particular, suggests that the principles of discant extemporization played a considerable role in his conception of musical works. Just as chansons could be used to provide the framework for extended mass settings, so might the discant-based frameworks used to construct Caron's sacred polyphony have migrated to the secular realm.

Owing to the construction of the tenor statements in the Naples complex, however, it is difficult to compare them with the extended practices observable in Caron's oeuvre. The systematized qualities of the six mass tenors introduce no variables within the cantus firmus presentations; each tenor adheres to a predetermined rhythmic plan consisting of little more than breves and semibreves. So although melodic substitutions can be extensive and often make shared structural gestures difficult to identify, the elaborations always occur against identical tenor structures. This surely explains why many of the elaborations have been misinterpreted as borrowings or demonstrative of thematic unity: by accommodating an identical tenor, they regularly have little need to be varied in themselves.

Fig. 3.8 Caron, *Missa Clemens et benigna,* Qui tollis, mm. 79–85.

In some instances this lack of variation is quite astonishing. For all the skill with which discant-based frameworks are manipulated within the set, occasional passages are repeated almost verbatim. This occurs to a limited extent in Caron's masses. The masses *L'homme armé* and *Jesus autem transiens* contain a handful of almost exact repetitions, which are typically no more than two measures in length (the main differences between the passages being rhythmic alterations, split notes, etc.). As we might expect, *Missa Clemens et benigna* contains several repetitions, of which the passage given in Figure 3.8 is found most frequently.

Some of the near-exact repetitions in the Naples masses are comparable in detail to this passage. What is different about several of the examples, however, is that they are far more protracted. In Mass I, for instance, the opening statements of the cantus firmus in the Gloria are accompanied by contrapuntal material that is taken over almost verbatim in the Credo (see Figure 3.9). As in Figure 3.1, the melodic twists and turns that distinguish the passages differ in a manner perhaps comparable to the reworkings of the London leaf:[44] one melodic pattern can be easily substituted for another with little resulting impact upon the overall harmonic foundations or polyphony. But in this case the substitutions are less pronounced than in passages of moderate embellishment (e.g., Figure 3.1). Most telling in this regard is the fact that the variants typically occur within the context of counterpoint in which one or more of the other discanting voices contains melodic material common to both frameworks at that point.

Many instances are known in which material from one or more non-cantus-firmus-bearing voice parts is repeated (in whole or in part) during the duration of a work. Most typically this occurs in mass settings in which two or more voices have been borrowed from a polyphonic model. Recent

Fig. 3.9 Anonymous, *Missa L' homme armé* (Mass I) (a) Et in terra, mm. 16 – 27; (b) Patrem, mm. 16–27; transcriptions after *Six Anonymous L'homme armé Masses in Naples, Biblioteca Nazionale, MS VI E 40,* ed. Judith Cohen, CMM 85 (Neuhausen-Stuttgart: Hänssler/AIM, 1981), 8–9 (a); 16–17 (b); used with permission.

research has suggested that the practice was far more widespread than first imagined. Before Josquin and Obrecht's generation, Ockeghem, Barbignant, Basiron, Bedyngham, Caron, Heyns, Le Rouge, Faugues, and Vincenet are all attributed with masses that attest to the tradition.[45] Martini's reworking of a German tenorlied in his *Missa In feuers hitz* provides perhaps the most interesting counterpart to the process described in the first Naples mass and found (to a lesser degree) in others of the complex.[46] Although based primarily on the tenor, extensive borrowings from the superius of *In feuers hitz* are found throughout the work. In a process similar to that found in the Naples masses, the paraphrases in the superius are treated more freely than in the tenor; as a rule, the quotations typically occur at the start of the main mass sections, with the result that the latter portion of the movements consist primarily of new material. Martini's treatment of paraphrased or quoted material in *Missa In feuers hitz* is almost identical in principle to the longer reiterated passages in Naples. The positioning of the passages and greater degree of variation in the superius are strong pointers to a shared compositional aesthetic.

The features that argue against such an interpretation, however, are surely more compelling. First, there is little stylistic distinction in the Naples masses between those passages containing ostensibly borrowed material and those of original counterpoint. Second, the lack of literal or even paraphrased repetitions of material from one or more contrapuntal voices in mass sections that quote different parts of the cantus firmus (or none at all) would suggest that the material is unlikely to have been borrowed from one or more polyphonic models. So if the repeated passages are not derived from a model, what is the significance of the material that is reiterated? More specifically, how might we understand the rationale underlying its presence or absence, length, and degree of elaboration? Ironically, it is the sheer variety that marks the treatment of such passages that is likely to provide the very clues needed to answer these questions.

As mentioned earlier, one can identify every imaginable degree of restatement and recomposition of material within each mass. The shared cantus-firmus-based passage of the Gloria and Credo discussed above and the short passages in Figures 3.2 and 3.3 might be imagined to occupy the far ends of a spectrum in which these two compositional processes are contrasted. Between restatement and recomposition lie those passages in which melodic gestures are not repeated per se but that nonetheless share the basic contours and rhythmic patterns necessary to make recognition of the framework relatively easy (Figure 3.1 stands as a fairly typical instance of this practice).

At the opposite end of the spectrum, then, stand those instances in which passages are extensively recomposed. In the Naples complex, corresponding passages of cantus firmus material can be found to possess almost identical

(a)

(b)

Fig. 3.10 Anonymous, *Missa L' homme armé* (Mass II) (a) Qui tollis, mm. 45–50; (b) Crucifixus, mm. 45–50; transcriptions after *Six Anonymous L'homme armé Masses in Naples, Biblioteca Nazionale, MS VI E 40,* ed. Judith Cohen, CMM 85 (Neuhausen-Stuttgart: Hänssler/AIM, 1981), 61 (a); 73 (b); used with permission.

structural foundations while nonetheless sharing little common melodic material. The Qui tollis and Crucifixus sections of Mass II shown in Figure 3.10, for instance, are clearly quite unlike each other in surface detail but nonetheless are surely based on a shared harmonic framework. Unlike Figure 3.9, the two passages do not share a common phrase structure despite a common tenor: the superius "directs" its accompanying voices from harmony to harmony via the sheer strength of its voice leading and anticipation of the next tenor pitch, rather than any preconceived blueprint based on cumulative phrase length. In other words, the florid melodic and rhythmic embellishments serve to mask an almost identical harmonic outline. This is surely not a reworking but rather a recomposition of a standardized series of vertical sonorities routinely applied to a tenor progression (in this case not a stepwise progression but an alternation between tonic and dominant pitches). What is perhaps most impressive about this example is the fact that melodic types and phrase patterns do not need to be shared in order to sustain the underlying structure. This surely demonstrates the degree to which melodic *varietas* could disguise otherwise standardized contrapuntal progressions.

As an indirect comparison, this example might be considered in light of the more extensive paraphrase techniques found in some of Isaac's mass settings based on polyphonic models. In these works, although the melodic and rhythmic structure of the model can still be identified at least to a limited degree, the quotations can be so complex that "the distinction between paraphrase and recomposition is often blurred."[47]

On the other hand, of course, many of the progressions found in the Naples masses are easily recognizable. As we have seen, it is primarily these that have been presented as evidence of the utilization of allusion, quotation, or paraphrase techniques in the Naples complex and related mass settings. The examples discussed earlier, however, demonstrate that these passages merely represent one extreme of a technique through which established contrapuntal progressions could be manipulated to varying degrees. The Naples set is of particular importance in this respect, for as a *complex* of mass settings, it contains a series of frameworks that are subject to a wide range of treatments. Some gestures are repeated with little alteration, while others—such as Figure 3.10—are subject to extensive melodic and rhythmic modifications.

With respect to issues of authorship, the use of comparable melodic frameworks in both the masses of Caron and those of the Naples set may suggest that the works are indeed associated. While many of the frameworks highlighted may be found in other mass settings, those of Caron and Naples are distinguished by the degree to which their musical structures are permeated by them. As both Giller and Reynolds showed, the high number of harmonic and melodic comparisons that can be made between the works is certainly intriguing.[48] But intrigue turns to frustration if we try to use these structures as tools for making attributions. The stylistic nature of the Naples masses means that there is little value in trying to attribute them on the basis of "borrowings" alone. Fundamentally, the masses are likely to be associated with Caron's works only via their utilization of shared discant-based frameworks—of which no individual is likely to have been the author. That is to say, it would be more appropriate to think of these works as being associated simply by closely related extemporization practices.

Whether these works owe their correspondences to a *single* localized or institutionalized practice is difficult to assess, as little is known about exactly how much discant traditions varied regionally. Certain stylistic features can point to vague geographic origins for a work, but considering that few polyphonic masses have been convincingly associated with specific liturgies or institutions, the task of making these designations more specific may well be an impossible one. But if we are to accept that extemporized polyphony and its notated counterpart did indeed share a close relationship, then it is necessary to accept that the features of these works may have been shaped as much by their institutional "obligations" as by the preferences of any one individual. Several compositional traditions have already been identified, for instance, in which liturgical links were clearly maintained more rigorously than in others. Mass Proper cycles such as those in the Trent manuscripts, for instance, display a stronger sense of liturgical unity than of musical unity.[49] Masses *de beata virgine* are similar, although here liturgical function is emphasized via the use of an established set of chant melodies as cantus firmi.[50] In a more general sense, Jennifer Bloxam has also demonstrated that localized chant traditions often played a crucial role with regard to the precompositional strategies of many northern composers.[51]

But while such research has highlighted the importance of localized chant traditions and liturgical practices in polyphonic mass cycles, the unquestionable influence of institutionalized *compositional* traditions upon sacred polyphony remains largely unexplored.[52] Mary Natvig recently concluded that "geographical tastes or preferences might account for" certain idiosyncrasies in a Magnificat tentatively attributed to Busnoys,[53] but the fact that her speculations are no more specific reflects the current state of knowledge in this area. The task of connecting specific compositions with particular locales and institutions (particularly when the works are of unknown origin) constitutes a formidable challenge. Yet our difficulties in this respect sit decidedly uncomfortably with contemporary observations regarding the aurally distinct qualities of many discant traditions. As Rob Wegman has clearly observed, "counterpoint diversified into regional and national 'manners,' even distinct techniques."[54] The question we must answer in this respect, however, is whether liturgical compositions of this period were any less distinct in terms of regional character.

This question is an understandably difficult one. With regard to the Naples masses it needs to be considered in light of a compositional aesthetic that began to emerge in the 1470s. In Rob Wegman's words, "to write a mass inevitably meant to define one's position. From the very moment a composer set out to design a new cycle, he was forced to make choices that would betray his preference and backgrounds."[55] While this vision is undoubtedly viable for some masses, it seems quite inappropriate for those discussed in this study. On one hand, Wegman's assessment is entirely valid. After all, it was during the 1470s that personal styles begin to be objectified. This was the decade, for instance, in which Martini was appointed *cantadore compositore* at the court of Ferrara[56] and when Tinctoris famously named contemporary composers and claimed to imitate "their admirable style of composition."[57] On the other hand, the articulation of such values was a relatively new phenomenon and—one suspects—so was the practice it referred to. It is in this light that we should consider both Caron's masses and the Naples complex. These works surely owe more to the particular style of an unidentified institution than to any individual (whether identified or anonymous). Indeed, the compositional frameworks identified in the masses suggest that the close relationship shared between extemporized and notated polyphony was by no means severed by the advent of the "composer."

Notes

1. For discussions of this issue, see Lewis Lockwood, "On 'Parody' as Term and Concept in 16th-Century Music," in Jan LaRue et al., eds., *Aspects of Medieval and Renaissance Music: A Birthday Offering to Gustave Reese* (New York: W. W. Norton, 1966; reprint, New York: Pendragon, 1978), 560–75; Howard Mayer Brown, "Emulation, Competition, and Homage: Imitation and Theories of Imitation in the Renaissance," *JAMS* 35 (1982): 1–48; Leeman L. Perkins, "The L'Homme Armé Masses of Busnoys and Okeghem: A Comparison," *Journal of Musicology* 3 (1984): 363–96; Rob C. Wegman, "Another 'Imitation' of Busnoys's *Missa L'Homme armé*–and Some Observations on *Imitatio* in Renaissance Music," *JRMA* 114 (1989): 189–202. Honey Meconi specifically addressed the issue of terminology in "Does *Imitatio* Exist?" *Journal of Musicology* 12 (1994): 152–78.
2. The term was first defined in Christopher A. Reynolds, "The Counterpoint of Allusion in Fifteenth-Century Masses," *JAMS* 45 (1992): 228–60; and Reynolds, *Papal Patronage and the Music of St. Peter's, 1380–1513* (Berkeley, Los Angeles, and London: University of California Press, 1995).
3. Rob C. Wegman, "From Maker to Composer: Improvisation and Musical Authorship in the Low Countries, 1450–1500," *JAMS* 49 (1996): 409–79, at 444. In a more general sense, see Richard L. Crocker, "Discant, Counterpoint, and Harmony," *JAMS* 15 (1962): 1–21; and Ernest T. Ferand, "Improvised Vocal Counterpoint in the Late Renaissance and Early Baroque," *Annales musicologiques* 4 (1956): 129–74, at 139–47.
4. See in particular Howard Mayer Brown, "Improvised Ornamentation in the Fifteenth-Century Chanson," *Quadrivium* 12 (1971): 238–58. David Fallows places greater emphasis on the role of the composer with regard to heavily embellished versions of polyphonic songs. He considers these works to be actual "recompositions" rather than embellished versions that simply "reflect performance practice of the time." See Fallows, "Embellishment and Urtext in the Fifteenth-Century Song Repertories," *Bäsler Jahrbuch für historische Musikpraxis* 14 (1990): 59–85.
5. The earliest embellishment manual is Sylvestro di Ganassi, *Opera intitulata Fontegara* (Venice: author, 1535). For a comprehensive list of such works, see Ernest T. Ferand, "Didactic Embellishment Literature in the Late Renaissance: A Survey of Sources," in Jan LaRue et al., eds., *Aspects of Medieval and Renaissance Music: A Birthday Offering to Gustave Reese* (New York: W. W. Norton, 1966; reprint ed., New York: Pendragon, 1978), 154–72.
6. Although the antiphon is likely to date from earlier in the century, the elaborated version of the *Da pacem* antiphon was copied between 1474 and 1476 by the main scribe of the manuscript, Thomas Ausquier; see Reynolds, *Papal Patronage,* 83, 89–98.
7. On the use of this antiphon in the Roman tradition, see Reynolds, *Papal Patronage,* 83. Calls for embellishment of this type, but with regard to chant melodies, are found in the statutes of other Italian cities. An ordo written for the cathedral of Siena in 1215, for example, calls for the embellishment of mass and Office chants; see Kurt von Fischer, "Die Rolle der Mehrstimmigkeit am Dome von Siena zu Beginn des 13. Jahrhunderts," *Archiv für Musikwissenschaft* 18 (1961): 167–82. Similar practices seem to have occurred in Lucca and Toledo; for the former, see Agostino Ziino, "Polifonia nella cattedrale di Lucca durante il XIII secolo," *Acta musicologica* 47 (1975): 16–30; for the latter, see Karl-Werner Gümpel, "Cantus

Eugenianus–Cantus melodicus," in Daniel Heartz and Bonnie Wade, eds., *International Musicological Society: Report of the Twelfth Congress, Berkeley 1977* (Kassel, Basel, and London: Bärenreiter, for the American Musicological Society, 1981), 407–13. Documented from 1391 though the eighteenth century, the practice in France was known as *machicotage;* see Mary Berry [Mother Thomas More], "The Performance of Plainsong in the Later Middle Ages and the Sixteenth Century," *Proceedings of the Royal Musical Association* 92 (1965–66): 121–34; and also Don Harrán, *In Defense of Music: The Case for Music as Argued by a Singer and Scholar of the Late Fifteenth Century* (Lincoln: University of Nebraska Press, 1989), 43–64.

8. For a fascinating discussion of the relationship between written and performed music, see Anne Stone, "Glimpses of the Unwritten Tradition in Some *Ars subtilior* Works," *Musica Disciplina* 50 (1996): 59–93.
9. See in particular those fragments discussed in Fallows, "Embellishment and Urtext," 66–70; Paula Higgins, "Music and Musicians at the Sainte-Chapelle of the Bourges Palace, 1405–1515," in Angelo Pompilio et al., eds., *Atti del XIV Congresso della Società Internazionale di Musicologia: Trasmissione e recezione delle forme di cultura musicale* (Turin: EDT, 1990), 3:689–701; and Bonnie J. Blackburn "A Lost Guide to Tinctoris's Teachings Recovered," *Early Music History* 1 (1981): 29–116.
10. A feature of the chant embellishments in LonBL 70516, noted in Fallows, "Embellishment and Urtext," 69.
11. See Ferand, "Didactic Embellishment Literature," 155.
12. Higgins, "Music and Musicians at the Sainte-Chapelle," 691.
13. For a detailed and thoughtful study of discant practices, see Wegman, "From Maker to Composer."
14. For a general study of these trends, see Margaret Bent, "Some Criteria for Establishing Relationships between Sources of Late-Medieval Polyphony," in Iain Fenlon, ed., *Music in Medieval and Early Modern Europe: Patronage, Sources and Texts* (Cambridge: Cambridge University Press, 1981), 295–318. For a particularly interesting study on manuscript copying techniques and their impact upon notated polyphony, see Rob C. Wegman, "*Miserere supplicanti Dufay:* The Creation and Transmission of Guillaume Dufay's *Missa Ave regina celorum,*" *Journal of Musicology* 13 (1995): 18–54.
15. Vicente Lusitano, *Introdutione facilissima et novissima* (Rome: Antonio Blado, 1553), fol. Dv, cited in Ferand, "Improvised Vocal Counterpoint," 150–51.
16. Leo Treitler, "Homer and Gregory: The Transmission of Epic Poetry and Plainchant," *Musical Quarterly* 60 (1974): 333–72, at 345.
17. Thus far, major studies include those by Ludwig and Geering for German-speaking regions and by Gomez Muntané and Anglès for Catalonia; see Friedrich Ludwig, *Repertorium organorum recentioris et motetorum vetustissimi stili, I: Catalogue raisonné der Quellen,* 2 vols. (Vol. 1, Halle: M. Niemeyer, 1910; reprint ed., Brooklyn: Institute of Mediaeval Music, 1964; Vol. 2, n.p.: Institute of Mediaeval Music, 1978); Arnold Geering, *Die Organa und mehrstimmigen Conductus in den Handschriften des deutschen Sprachgebietes vom 13. bis 16. Jahrhundert* (Bern: P. Haupt, 1952); Higini Anglès, *La música a Catalunya fins al segle XIII* (Barcelona: Institut d'estudis Catalans/Biblioteca de Catalunya, 1935); Maria Gomez Muntané, "El ars antigua en Cataluña," *Revista de musicologia* 2 (1979): 279–84. See also Kurt von Fischer and Max Lütolf, eds., *Handschriften mit mehrstimmiger Musik des 14., 15. und 16. Jahrhunderts,* 2 vols., RISM B IV$^{3-4}$ (Munich-Duisburg: G. Henle, 1972). Much-needed source lists have been provided for Italy (Gallo), Austria (Flotzinger), England (Reaney, Bowers and Wathey), and Scandinavia (Bergsagel); see F. Alberto Gallo, "The Practice of *Cantus Planus Binatim* in Italy from the Beginning of the 14th to the Beginning of the 16th Century," in Cesare Corsi and Pierluigi Petrobelli, eds., *Le Polifonie primitive in Friulie in Europa: Atti del congresso internazionale Cividale del Friuli, 22–24 agosto 1980* (Rome: Edizione Torre d'Orfeo, 1989), 13–30; Rudolf Flotzinger, "Non-mensural Sacred Polyphony (*Discantus*) in Medieval Austria," in Cesare Corsi and Pierluigi Petrobelli, eds., *Le Polifonie primitive in Friuli e in Europa: Atti del congresso internazionale Cividale del Friuli, 22–24 agosto 1980* (Rome: Edizione Torre d' Orfeo, 1989), 43–61; John Bergsagel, "The Practice of *Cantus Planus Binatim* in Scandinavia in the 12th to 16th Centuries," in Cesare Corsi and Pierluigi Petrobelli, eds., *Le Polifonie primitive in Friuli e in Europa: Atti del congresso internazionale Cividale del Friuli, 22–24 agosto 1980* (Rome: Edizione Torre d' Orfeo, 1989), 63–82; Gilbert Reaney, "Some Little-Known Sources of Medieval Polyphony in England," *Musica Disciplina* 15 (1961): 15–26; Roger Bowers and

Andrew Wathey, "New Sources of English Fourteenth- and Fifteenth-Century Polyphony," *Early Music History* 3 (1983): 123–73.

18. Many examples are found throughout the secondary literature. For a general evaluation with regard to borrowing techniques, see Brown, "Emulation, Competition, and Homage"; Reynolds, "The Counterpoint of Allusion"; and Reynolds, *Papal Patronage.*
19. Reynolds, *Papal Patronage,* 141–50.
20. Margaret Bent, "*Resfacta* and *Cantare Super Librum,*" *JAMS* 36 (1983): 371–91, at 389–91.
21. Wegman, "From Maker to Composer," 432.
22. Johannes Tinctoris, *Liber de arte contrapuncti,* Book II, Chapter xx; for this passage and a translation, see Bonnie J. Blackburn, "On Compositional Process in the Fifteenth Century," *JAMS* 40 (1987): 210–84, at 248–49.
23. Crocker, "Discant, Counterpoint, and Harmony," 10, 16.
24. The masses were first reported by Dragan Plamenac, "La Chanson de L'Homme armé et MS. VI. E. 40 de la Bibl. Nationale de Naples," in *Annales de la Fédération archéologique et historique de Belgique, Congrès jubilaire* (Bruges, 1925), 229–30. Two editions of the masses exist: Laurence Feininger, ed., *Missa prima [–VI] super L'homme armé, auctore ignoto, secundum unicum fontem mutilem, scilicet Codicem Neapolitanum Ms. VI–E.40,* Monumenta Polyphoniae Liturgicae Sanctae Ecclesiae Romanae I/3 (Rome: Societas Universalis Sanctae Ceciliae, 1957–74); and Judith Cohen, ed., *Six Anonymous L'homme armé Masses in Naples, Biblioteca Nazionale, MS VI E 40,* CMM 85 (Neuhausen-Stuttgart: Hänssler/AIM, 1981).
25. Richard Taruskin, "Antoine Busnoys and the *L'homme armé* Tradition," *JAMS* 39 (1986): 255–93, at 279–80; Don Giller, "The Naples L'Homme Armé Masses and Caron: A Study in Musical Relationships," *Current Musicology* 32 (1981): 7–28; Judith Cohen, "Munus ab ignoto," *Studia musicologica Academiae scientiarum hungaricae* 22 (1980): 187–204, at 198–99; Barbara Helen Haagh, "Communication," *JAMS* 40 (1987): 139–43, at 142; Reynolds, *Papal Patronage,* 250–52, 278.
26. Leo Treitler, "'Centonate' Chant: *Übles Flickwerk* or *E pluribus unus*?" *JAMS* 28 (1975): 1–23, at 1.
27. Cohen, "Munus ab ignoto," 198–99.
28. Giller, "The Naples L'Homme Armé Masses."
29. Cohen, "Munus ab ignoto," 42–43; Giller, "The Naples L'Homme Armé Masses," 12–21; Reynolds, *Papal Patronage,* 250–52, 267, 278.
30. Peter Jeffery, *Re-Envisioning Past Musical Cultures: Ethnomusicology in the Study of Gregorian Chant* (Chicago: University of Chicago Press, 1992), 87–90.
31. Wegman, "From Maker to Composer," 442.
32. All passages from Caron's works are taken from Firmin Caron, *Opera omnia,* ed. Jenny Hodgson, CMM 111 (Middleton, Wisc.: AIM, forthcoming).
33. Reynolds, *Papal Patronage,* 266–69.
34. Meconi argues that "much polyphonic borrowing is clearly audible" for the reason that the quotations could be more easily recognized by listeners; see Meconi, "Does *Imitatio* Exist?" 175.
35. Giller, "The Naples L'Homme Armé Masses," 9–10.
36. Treitler, "Homer and Gregory," 345, 352–53, 362–66.
37. For a detailed discussion of the role of the *tenorist* and related bibliography, see Wegman, "From Maker to Composer," 444–49.
38. Giller, "The Naples L'Homme Armé Masses," 11–12.
39. In this respect, see in particular the chapter "Musical Connoisseurship" in Reynolds, *Papal Patronage,* 141–71.
40. See Hans Schmidt, *Zum Formelhaften Aufbau byzantinischer Kanones* (Wiesbaden: Breitkopf & Härtel, 1979); and Nina K. Ulff-Møller, "The Connection between Melodic Formulas and Stereotype Text Phrases in Slavonic Stichera," *Cahiers de l'Institut du Moyen-âge grec et latin, Université de Copenhague* 54 (1986): 49–60.
41. Jeffery, *Re-Envisioning Past Musical Cultures,* 91.
42. Reynolds, *Papal Patronage,* 275–77.
43. Ibid., 275.
44. LonBL 70516, fol. 79. This single parchment sheet consists of a series of embellishment patterns over eight three-note tenors. David Fallows has proposed that the sheet "is a

demonstration of how to improvise over a chant tenor." For more information, see Fallows, "Embellishment and Urtext," 66–70.

45. For a more comprehensive list and bibliography, see Murray Steib, "Imitation and Elaboration: The Use of Borrowed Material in Masses from the Late Fifteenth Century," (Ph.D dissertation, University of Chicago, 1992), 267–328.
46. For this and what follows, see J. Peter Burkholder, "Johannes Martini and the Imitation Mass of the Late Fifteenth Century," *JAMS* 38 (1985): 470–523, at 485.
47. Murray Steib, "A Composer Looks at His Model: Polyphonic Borrowing in Masses from the Late Fifteenth Century," *TVNM* 46 (1996): 5–41, at 14.
48. Giller, "The Naples L'Homme Armé Masses"; Reynolds, *Papal Patronage.*
49. See in particular Alejandro Enrique Planchart, "Guillaume Du Fay's Benefices and His Relationship to the Court of Burgundy," *Early Music History* 8 (1988): 117–71; Rebecca Gerber, "Dufay's Style and the Question of Cyclic Unity in the Trent 88 Mass Proper Cycles," in Peter Wright, ed., *I codici musicali trentini: Nuove scoperte e nuovi orientamenti della ricerca* (Trent: Provincia autonoma di Trento, Servizi[o] beni librari e archivistici, 1986), 107–19; and David Fallows, *Dufay* (London: J. M. Dent, 1982), 188–91.
50. Gustave Reese, "The Polyphonic 'Missa de Beata Virgine' as a Genre: The Background of Josquin's Lady Mass," in *JosCon,* 589–98, at 593. The choice of standardized chants was still localized during the fourteenth century; see Anne Walters Robertson, "The Mass of Guillaume de Machaut in the Cathedral of Reims," in Thomas Forrest Kelly, ed., *Plainsong in the Age of Polyphony* (Cambridge and New York: Cambridge University Press, 1992), 100–39, at 105–06.
51. M. Jennifer Bloxam, "In Praise of Spurious Saints: The *Missae Floruit egregiis* by Pipelare and La Rue," *JAMS* 44 (1991): 163–220; Bloxam, "Sacred Polyphony and Local Traditions of Liturgy and Plainsong: Reflections on Music by Jacob Obrecht," in Thomas Forrest Kelly, ed., *Plainsong in the Age of Polyphony* (Cambridge and New York: Cambridge University Press, 1992), 140–77; and Bloxam, "On the Origins, Contexts, and Implications of Busnoys's Plainsong Cantus Firmi: Some Preliminary Remarks," in Paula Higgins, ed., *Antoine Busnoys: Method, Meaning, and Context in Late Medieval Music* (Oxford: Clarendon, 1999), 71–88. See also Mitchell P. Brauner, "Traditions in the Repertory of the Papal Choir in the Fifteenth and Sixteenth Centuries," in Richard Sherr, ed., *Papal Music and Musicians in Late Medieval and Renaissance Rome* (Oxford: Clarendon, in association with the Library of Congress, Washington, D.C., 1998), 167–74; and Adalbert Roth, "Liturgical (and Paraliturgical) Music in the Papal Chapel towards the End of the Fifteenth Century: A Repertory in Embryo," in Richard Sherr, ed., *Papal Music and Musicians in Late Medieval and Renaissance Rome* (Oxford: Clarendon, in association with the Library of Congress, Washington, D.C., 1998), 125–37.
52. Two short studies pertaining to Josquin's motets are of some interest in this respect; see Jacquelyn A. Mattfeld, "Some Relationships between Texts and Cantus Firmi in the Liturgical Motets of Josquin des Pres," *JAMS* 14 (1961): 159–83; and Jeremy Noble, "The Function of Josquin's Motets," *TVNM* 35 (1985): 9–31. Chanson literature is more varied is its investigation of influences; for a particularly interesting study and bibliography, see Walter H. Kemp, *Burgundian Court Song in the Time of Binchois: The Anonymous Chansons of El Escorial, MS V.III.24* (Oxford: Clarendon, 1990).
53. Mary Natvig, "The Magnificat Group of Antoine Busnoys: Aspects of Style and Attribution," in Paula Higgins, ed., *Antoine Busnoys: Method, Meaning, and Context in Late Medieval Music* (Oxford: Clarendon, 1999), 257–76, at 275.
54. Wegman, "From Maker to Composer," 421.
55. Rob C. Wegman, "The Anonymous Mass *D'Ung aultre amer:* A Late Fifteenth-Century Experiment," *Musical Quarterly* 74 (1990): 566–94, at 567.
56. Lewis Lockwood, *Music in Renaissance Ferrara, 1400–1505: The Creation of a Musical Center in the Fifteenth Century,* Studies in the History of Music 2 (Oxford: Clarendon; New York: Oxford University Press; Cambridge, Mass.: Harvard University Press, 1984), 154–55.
57. Johannes Tinctoris, *The Art of Counterpoint (Liber de arte contrapuncti),* trans. Albert Seay, Musicological Studies and Documents 5 (n.p.: AIM, 1961), 15.

4
Interpreting and Dating Josquin's *Missa Hercules dux ferrariae*

CHRISTOPHER REYNOLDS

In two earlier studies of motivic allusion in the fifteenth century, I found it useful to describe allusion—whether literary or musical—as a rhetorical technique.[1] In so doing I attempted to draw parallels between the practices long documented in poetry and oratory and those evident in music. More recently, in studies of nineteenth-century repertoires, I argue that Romantic theories of symbols and symbolization had supplanted rhetoric as a means of theorizing the same kinds of motivic references. In both centuries it was common for borrowed ideas to be varied. As Howard Brown explained it for music, when chanson composers emulated each other, earlier material was not incorporated into a new context literally, but was "reshaped and rearranged into new musical phrases."[2] This is precisely the formula for citing earlier works that the humanist Gasparino Barzizza prescribed in the early fifteenth century: "all good literary imitation comes from adding, subtracting, altering, transferring, or renewing."[3] In the nineteenth century composers had incentive to disguise musical debts both because the demands for artistic originality required it and because artistic symbols needed to veil their references in order for the references to achieve the status of symbols.[4]

Yet rhetoric and theories of symbols are constructs for explaining a form of communication that has deeper roots. Whether from the fifteenth or nineteenth century, whether musical or poetic, allusion was fundamentally a form of play. Both the English word *allude,* which comes directly from the Latin *alludere* (to play), and the German word *Anspiel,* which similarly derives from *spiel* (to play), mean "to play on." Allusion plays on words in literature, on images in art, and on any aspect of music, but especially motives, with or without text. In his inspired study of the human impulse to play, *Homo Ludens,* Johan Huizinga referred to language when he asserted that "behind every abstract expression there lie the boldest of metaphors, and every metaphor is a play on words."[5] But as an activity based on the "imagination of reality (i.e., its conversion into images)," allusion is directly linked to symbolization, and thus it is not only accessible to all arts but common to all eras and all forms of communication.[6] Those who deny the possibility of motivic allusion argue that because there are so few notes, motivic similarities are inevitable and therefore meaningless. However, this overestimates the possibilities enjoyed by artists and poets: artists were not prevented from investing colors or images with symbolism by the limited number of primary colors or images (and the fact that any one color or image had multiple symbolic associations), and poets usually created allusions not with individual words but with groupings of words that were constrained both by syntactical conventions and by the limited number of meaningful rhymes.

Notions of play are relevant to discussions of how composers transformed motives to conceal relationships, how composers represented themselves and others by means of musical motives (e.g., B-A-C-H and "Hercules dux ferrariae"), and why composers alluded and how they construed this act. In the fifteenth century allusion in any artistic medium provided multiple challenges for the artistic creator and the listener/reader. The intricate and complementary acts of concealment and interpretation brought artist and audience (small or large) together in an intellectual game of symbolic hide-and-seek.

Aside from the rhetorical tradition of altering a model, several traits make the fifteenth century (and the first half of the sixteenth) a period particularly fertile for creating and reading motivic allusions:

- It was an age that valued wit and artifice, as evident in innumerable literary works that quote and allude, in musical works that refer to their own composers (e.g., Busnoys, Du Fay, Compère) or chansons that borrow musically and textually from another chanson (e.g., Joye's *Mercy, mon dueil* and the anonymous *Mon cuer de dueil*),[7] and in the many displays of contrapuntal or mensural ingenuity and intricate canons.
- There was a common international repertoire, with composers in the north and south contributing to the same pool of secular and sacred works.

- It was a vocal era, which both places greater limitations on the variety of motives and means that motives could have textual associations.
- It was—as usual before the twentieth century—a time when musicians improvised. The resultant formulaic character of compositional styles meant that music was routinely constructed in terms of small units, whether motives, ornamental extensions of motives, or cadential figures. Opportunities for allusion, both those intended by a composer and those perceived by a performer or listener, arose because of the frequent resemblances. That for most of the past century the only shared repertoire has been the inherited canon, with modernist musical styles neither vocal nor improvisatory—excepting in all respects jazz—helps to explain modern difficulties in recognizing the allusive practices of earlier periods.

In this study I will examine motives for two distinct purposes. One is hermeneutic (to show how the allusions I posit can shape an interpretation of Josquin's mass), and the other is to assist in dating the mass by means of motivic resonances with other works that appear to be contemporaneous. It is not necessary to accept my arguments for allusion to entertain those for establishing a date: Josquin's mass has strong stylistic links to other works, and these have implications for arguments about when Josquin composed it.

* * *

However distinctive the ostinato of Josquin's *Missa Hercules dux ferrariae,* however ingenious the hexachordal derivation of this famous motive from the vowels of Ercole d'Este's name, and however influential this mass for Lupus Hellinck, Willaert, Gombert, Morales, Cipriano de Rore, and others who honored later rulers with similar masses,[8] Josquin's *soggetto cavato* may not have been entirely original; rather, he may have adapted it from a phrase in Walter Frye's *Missa Nobilis et pulchra.* While it is indisputable that once Josquin chose to derive Ercole's motto from the words "Hercules dux ferrariae" the notes were prescribed, my contention in this essay is that Josquin's decision to honor Ercole with a *soggetto cavato* in the first place (rather than with some more traditional cantus firmus), as well as his choice of those precise words (rather than others available to him from commemorative coins and medals), were made because this combination of vowels yielded a motive with layers of allusive meaning. That no motive is unique, even a motive contrived to express the letters of a patron's name, title, and city, is a point that should be self-evident.[9] The Hercules motto, distinctive as it is, compares closely to two others from the period, the motive that Frye used to set the text "ex Maria virgine" in *Missa Nobilis et pulchra,* and the opening motive of an anonymous *Salve regis mater* (probably by Marbriano

Fig. 4.1 (a) Josquin des Prez, *Missa Hercules dux ferrariae,* Kyrie I; (b) Walter Frye, *Missa Nobilis et pulchra,* Patrem; (c) anonymous [De Orto?], *Salve regis mater.*

de Orto) written for the coronation of Pope Alexander VI on August 26, 1492 (see Figure 4.1).[10]

But before examining the possible significance of these similarities, I would like to begin by discussing an extended series of motivic and structural similarities between *Missa Hercules* and another of Josquin's ostinato masses, *Missa Faisant regrets.* The latter mass has a short tenor motive taken from Walter Frye's chanson *Tout a par moy.* Although the motive is only four notes long and extracted not from the beginning of the chanson but from the middle, the derivation of the mass ostinato from the chanson has never been in doubt because Josquin appropriated Frye's words as well as his notes. Had the mass instead been called *Missa Fa-re-mi-re* or named after some dedicatee whose name accommodated those syllables (for example, W*a*l-t*e*r Fr*y*-*e*), the connection to *Tout a par moy* would be unrecognizable.

Fig. 4.2 Josquin (a) *Missa Hercules dux ferrariae,* Patrem; (b) *Missa Faisant regrets,* Patrem.

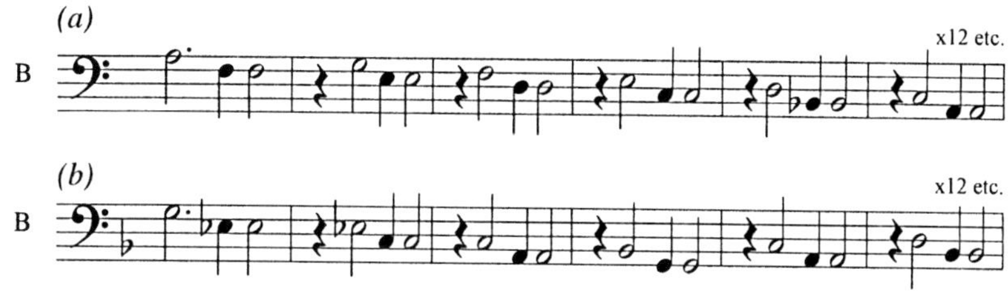

Fig. 4.3 Josquin (a) *Missa Hercules dux ferrariae,* Kyrie II; (b) *Missa Faisant regrets,* Sanctus.

The similarities of this mass with Josquin's *Hercules* mass include the treatment of the motto and the identity of various contrapuntal ideas. Unlike Josquin's other ostinato mass, the *Missa La sol fa re mi,* the tenors of the *Faisant regrets* and *Hercules* masses present only the motive at different transpositions and speeds.[11] At the start of the Patrems, Josquin introduces the chant melody of Credo I in the two upper voices (see Figure 4.2). In *Missa Hercules* the motive appears in imitation at the fifth above, in *Missa Faisant regrets* at the fifth below.[12] Equally close are the *Faisant regrets* Sanctus and the *Hercules* Kyrie II. The first measures of the bassus lines are juxtaposed in Figure 4.3. Both of these voices take this motive over from the superius and state it exactly twelve times.[13] Elsewhere, the *Faisant regrets* Pleni and the *Hercules* Osanna commence with similar imitative motives (see Figure 4.4). The Osannas of both masses are close structurally: they present

Fig. 4.4 Josquin (a) *Missa Hercules dux ferrariae,* Osanna; (b) *Missa Faisant regrets,* Pleni.

Fig. 4.5 Josquin (a) *Missa Hercules dux ferrariae,* Agnus I; (b) *Missa Faisant regrets,* Agnus II.

their respective ostinatos not just in the tenor but in overlapping imitative statements in the two lower voices.

Most strikingly, two motivic ideas in the Agnus are very nearly identical. The *Faisant regrets* Agnus II and the *Hercules* Agnus I begin with a cadential figure that shares pitch, rhythm, and the way in which it is immediately repeated (see Figure 4.5). Even the "re-fa-mi-re" continuation is the same. And Agnus I of *Missa Hercules,* like Agnus III of *Missa Faisant regrets,* has a repetitive "mi-fa-mi" motive (see Figure 4.6). These last two movements also present the principal motive in retrograde. In *Missa Faisant regrets* Josquin quotes from the first phrase of the chanson in the upper three voices. But since Frye set the text "Tout a par moy" to a motive—"re-re-mi-re-fa"—that

Fig. 4.6 Josquin (a) *Missa Hercules dux ferrariae,* Agnus I; mm. 20–25; (b) *Missa Faisant regrets,* Agnus III, mm. 5–11.

Fig. 4.7 (a) Frye, *Tout a par moy;* (b) Josquin, *Missa Faisant regrets,* Agnus III.

is simply the retrograde of the "Faisant regrets" motive, the retrograde in the mass is disguised (see Figure 4.7). Josquin returns several times to his artful pairing of tenor and contra voices to end the mass. On the basis of just one of these similarities, Elders has suggested that "these two works originated at about the same time."[14] While I concur with his conclusion, especially in view of the many other ideas these two works share, for reasons given below I cannot agree that these works were written by the early 1480s.

Why should Josquin have linked these two masses so extensively, the one written for a Ferrarese duke, the other—as we will soon see—probably for Marguerite of Austria? Potentially underlying all of these musical parallels is a shared motivic debt to Walter Frye. The hexachordal motto of *Missa Hercules* arguably derives from an interior phrase of *Missa Nobilis et pulchra.* As Lewis Lockwood observed, the configuration Josquin used is not the only one possible. The "fa" syllable of "Ferrariae" could also have been rendered as "la," but in the conjunct form that Josquin devised, the motive resembles a standard formula for D-mode chants, such as that used in the first tenor notes of Frye's mass (see Figure 4.8a).[15] Based on the Responsory for St. Catherine of Alexandria, Frye's mass was evidently copied into the Burgundian manuscript BrusBR 5557 about 1468, the year of the wedding of Charles the Bold and Margaret of York. In his Agnus I Josquin prominently employs a contrapuntal motive that is particularly close to Frye's principal motive because it drops the reiteration of "re-ut" (see Figure 4.8b).[16] But much closer to the *Hercules* motive is Figure 4.1b, a phrase that appears just once in the mass, copied in BrusBR 5557 with the text "ex Maria virgine." Although there is not one reiteration of "re-ut" but two, the resemblance of this contratenor motive to the *Hercules* motto is striking and is enhanced by

Fig. 4.8 (a) Frye, *Missa Nobilis et pulchra,* Kyrie I; (b) Josquin, *Missa Hercules dux ferrariae,* Agnus I.

the accompanying counterpoint. Figure 4.9a shows both voices of the "ex Maria virgine" duet against a similar counterpoint in the Josquin Qui tollis (Figure 4.9b). Frye returns to this figure in Agnus II.

I am thus suggesting that Josquin's motive alludes to a motive from a contratenor part rather than a tenor. That composers could base a new composition on material drawn from voices other than the tenor of an earlier composition was acknowledged by Tinctoris in the 1470s in his *Proportionale musices:* "The contratenor, however, is rarely or never a primary part. If, nevertheless, above some precomposed melody we wish to add something, we make it primary." He then provides an example in which

Fig. 4.9 (a) Frye, *Missa Nobilis et pulchra,* Patrem; (b) Josquin, *Missa Hercules dux ferrariae,* Qui tollis.

Fig. 4.10 (a) Josquin, *Missa L'homme armé super voces musicales,* Agnus II; (b) Johannes Ockeghem, *Ma bouche rit,* transposed down a fifth.

two new voices are added to the contratenor of *Le serviteur.*[17] Josquin may have composed on a preexisting contratenor on at least one other occasion. The famous three-voice mensuration canon that he wrote for Agnus II of *Missa L'homme armé super voces musicales* is evidently based on a quotation of the first contra measures of Ockeghem's chanson *Ma bouche rit et mon cueur pleure* (see Figure 4.10). As the allusive referent for an Agnus, the full text of the chanson suitably pairs the plea for heavenly mercy with an extreme statement of mortal despair. Moreover, the existence of an allusion to Ockeghem in the context of a mensuration canon, along with the presence of other mensuration canons in the Kyrie and Benedictus, makes this mass an ideal work to have been copied into the Chigi Codex (VatC 234), with its focus on Ockeghem.

Ercole's motto also resembles the first motive of *Salve regis mater,* probably composed by Josquin's papal colleague de Orto in 1492 (see Figure 4.1c). Since Josquin sang in the Cappella Sistina at the time, he surely would have known this work. If Josquin's mass was written earlier, the motet could be adduced as corroborative evidence for the Marian allusion I have claimed for Josquin; it would thus constitute a second example of a Marian composition that incorporated this motive in a work designed to honor a leader (Pope Alexander VI). But if, as I will argue, Josquin's work was later, then there is no reason why Josquin could not have consciously alluded to both Frye and the papal motet. Again *Missa Faisant regrets* provides a precedent. Josquin may have gotten the idea of the "fa-re-mi-re" ostinato from Agricola's arrangement of Frye's *Tout a par moy* published by Petrucci in Canti C (1504^3). As Edward Lowinsky observed, Josquin demonstrably refers to both Frye and Agricola, since he quotes Frye's superius and Agricola does not.[18]

* * *

Biographical and paleographic considerations support these musical reasons for relating Ercole's motto and Frye's motive. First, the textual symbolism of this borrowing accords with what is known about Ercole's avowed religious beliefs; second, a relationship to *Missa Nobilis et pulchra* helps to explain curiosities in the manuscript transmission of *Missa Hercules;* and third, this would not be the only hexachordal-Marian composition that can be associated with Ercole.

A mass based on the "ex Maria virgine" motive from *Missa Nobilis et pulchra* would have pleased Ercole for obvious religious reasons. In an age when it was commonplace to recognize princes and kings with phrases such as "Defender of the Faith," "Most Christian Lord," and the like, it was not hyperbolic to call Ercole "*religiossisimo principe,*" as the writer Giovanni Sabadino did in his tract of 1497, *De triumphis religionis.*[19] One year before, Ercole had published his own religious booklet, a series of prayers to the Virgin Mary entitled *Corona Beatae Mariae Virginis.* The invocation "Ave mater dei, ora pro me" is the foundation of the *Corona,* with each word serving as the beginning of a separate prayer; thus the first word, "Ave," introduces the "Ave maris stella," and so forth. Ercole had his choir sing these prayers every week in a special Saturday office for the Virgin Mary.[20] Similarly, Ercole left provisions in his will for two Marian services in the cathedral every Saturday: a Virgin mass in the morning and a sung Rosary of the Blessed Virgin every evening. And for his tomb Ercole chose not the cathedral, dedicated to St. George, but a site before the high altar in Santa Maria degli Angeli. One of many public displays of his piety existed in the remarkable number of churches and convents he built, a demonstration of faith hailed as "unprecedented and unparalleled."[21]

Josquin's motto for Ercole would thus be directly comparable to fifteenth-century portraits in which the patron is depicted kneeling in front of the Virgin Mary. By citing *Missa Nobilis et pulchra,* Josquin symbolically joined Ercole's nobility with Mary's beauty. While this kind of linkage was routine for artists, it was ingenious musically for the richness of its allusions. It may also be relevant that Frye's mass was based on a chant for St. Catherine of Alexandria, a Virgin martyr. Throughout the Renaissance artists depicted the mystic marriage of St. Catherine to Christ as a symbolic dedication of her virginity. This story must therefore be considered as part of the allusive background on which Josquin could have drawn when he derived his tenor motto.

The identification of a musical and allusive source for Josquin's *soggetto* also helps to explain a widely noted curiosity about the manuscript tradition of *Missa Hercules,* namely, that in two Habsburg-Burgundian manuscripts the notes of the tenor are fitted with new aristocratic identities. In BrusBR 9126, which was prepared in the court scriptorium in 1505 for Archduke Philip the Fair, king of Castile, the mass was retitled *Philippus rex castiliae;* similarly, in another illustrated choirbook, JenaU 3 from ca. 1518–20, the mass was prepared for the German elector Frederick the Wise, and now entitled *Missa Fredericus dux saxoniae.* Rather than requesting a mass for Philip on the tenor "mi-mi-ut-re-fa-mi-mi-re," or for Frederick on the *soggetto* "re-re-mi-ut-ut-fa-sol-mi-re," whoever commissioned Josquin's mass to be fitted with new titles evidently had no doubt that the recipients of these deluxe codices would appreciate "their" honorific masses, whatever the disjunction between their names and the mass tenors.[22]

But why should this be so? Why should either Philip or Frederick be honored by masses whose underlying allusion was to Ercole d'Este? Ercole was regarded by some as the major Italian leader by the 1490s, and he was on excellent terms with the King of France, but he was simply not in the same international league as Philip.[23] In addition to his short-lived reign as king of Castile, Philip was archduke of Austria and titular duke of Burgundy. From his mother, Mary of Burgundy, he had at age four inherited the Netherlands, and he was the most likely successor to his father Maximilian I and thus presumptive heir to the title Holy Roman Emperor and all that went with it. When Ercole died in 1505, he was mourned in Italy; in comparison, Philip's death the next year sent political shock waves throughout Europe. To think that a mass for Ercole would flatter the new recipients, either one must assume that whoever commissioned the manuscripts was politically naive or that Philip and Frederick, two of the most musically literate princes of their day, would not have cared about the origins of "their" masses. Or, recognizing *Missa Hercules dux ferrariae* as a work that had honored Ercole with a Marian soggetto extracted from a mass sung in the chapel of Charles the Bold, it is easier to understand that Philip would have enjoyed both the Marian allusion and the musical link to a mass that, given its presence in BrusBR 5557, may have had an association with his grandfather, the last duke of Burgundy.

Frederick, for his part, rivaled Ercole in his religiosity and his Marian devotion. Recognized today for supporting Martin Luther, he was known in his own time for the sincerity with which he conducted his daily worship. Among several relevant projects, Frederick helped sponsor the publication of a Marian tract by the Augustinian monk Johann von Paltz; he had Cranach paint him on his knees before a portrait of the Virgin; and he commissioned, among other Marian artworks, Dürer's "Seven Sorrows and Seven Joys of Mary." This now dismantled altarpiece bore the Latin inscription "On account of you, I the Virgin demand the seven joys of which the joyful Prince Frederick speaks."[24]

This interpretation of *Missa Philipus rex castiliae* and *Missa Fredericus dux saxoniae* implies not only that Philip and Frederick would have recognized Josquin's debt to Frye's mass but that whoever ordered the retitled masses did as well. Evidence for such an awareness brings us back to Josquin's *Missa Faisant regrets.* In addition to similarities of motive and structure, both masses have in common a quirk of transmission: like *Missa Hercules,*

Missa Faisant regrets survives with an altered tenor text. In one of the three Habsburg-Burgundian manuscripts in which it was copied, it is named *Missa Elizabeth faisant regrets*, "Elizabeth" referring to the mother of Frederick the Wise. Tellingly, this retitled version is in JenaU 3, the manuscript that also contains *Missa Fredericus dux saxoniae*. Indeed, these two masses begin the codex: first the Elizabeth mass, then the Frederick. Because of this pairing, it seems all the more likely that whoever commissioned JenaU 3—Marguerite of Austria is a candidate—recognized the "ex Maria virgine" allusion of *Missa Hercules* and thus the Frye content of both masses.

The numerous correspondences between these two masses also increase the probability that Josquin originally intended the "Faisant regrets" motive to have a symbolic hexachordal meaning; to pose it as a question, if the "ex Maria virgine" motive was chosen as an emblem for Ercole d'Este, whom might "fa-re-mi-re" represent? Despite my passing suggestion of Walter Frye, a much more plausible dedicatee for *Missa Faisant regrets* has already been suggested by Herbert Kellman: Marguerite of Austria, spelling her name in the French fashion, "Mar-gue-ri-te." Kellman notes that "the third stanza of the poem 'Me fauldra-il,' often attributed to Marguerite," has a textual link with Frye's chanson *Tout a par moy*. The third stanza "begins with the line 'Parquoy conclus, seullette et a par moy . . .'" Kellman also observes a musical connection, since "in the anonymous setting of this text in BrusBR 228, the words 'a par moy' in the tenor fall on the motive fa-re-mi-re" according to his text underlay.[25] In addition to these arguments, it may be significant that Frye used this particular motive for the words "Faisant regrets," given the number of *Regrets* compositions that have been associated with Marguerite and manuscripts copied at her court.

These arguments have implications for the vexatious issue of determining when Josquin wrote *Missa Hercules dux ferrariae*. Proposed datings range widely from 1480 or 1481 to 1503, making it either one of his earliest known masses or a fully mature work. Lockwood initially favored the earlier of these, based on Lowinsky's theory that Josquin was probably a familiar of Ascanio Sforza during Sforza's residence for eighteen months in Ferrara in just these years.[26] In the meantime, the archival research of Paul and Lora Merkley has shown that Josquin was not likely to have served with Ascanio Sforza before 1484, thereby undermining arguments for an earlier date; they propose instead that Josquin composed the mass between 1484 and 1492, a span that includes dates suggested by Patrick Macey and also by Lowinsky, who proposed 1487, the year that Ercole visited Rome.[27] Late datings of ca. 1503, advanced by Helmuth Osthoff, Jeremy Noble, Joshua Rifkin, and Lockwood, coincide with Josquin's year-long employment in Ferrara.[28]

Certainly the mass stems from before June 1505, when Petrucci published it in Josquin's second book of masses, RISM J670. All of the manuscript sources are later than Petrucci, excepting the Milanese source Librone 3 (MilD 3). The Merkleys have attempted to date this manuscript between the years 1489 and 1499; most likely, in their view, it is from 1492.[29] Because this source transmits a version of the *Hercules* mass that has some variants not found in any other source, they argue that Josquin was responsible for the variants and that he therefore must have written the work during the years of his association with Milan, 1484–89. Yet *Missa Hercules* is one of three Josquin masses copied by a secondary scribe (first identified by Knud Jeppesen).[30] The Merkleys pass over the possibility that the secondary scribe added these works later. It was not unusual for principal scribes to leave blank folios for later additions; in VatSP B80 the last additions occurred nearly twenty-five years after the principal scribe had finished his copying.[31] Indeed, in the case of Josquin's *Missa L'homme armé sexti toni,* copied immediately before *Missa Hercules* by the same scribe, Rifkin has argued that this mass should be dated after 1502 because it appears to have been copied from Petrucci's first book of Josquin's masses, RISM J666 (an argument the Merkleys discount).[32]

The musical connections with *Missa Faisant regrets,* while they in no way confirm that the two masses were written in close chronological proximity, nevertheless are significant enough that the date of this mass must also be considered. Its earliest surviving copy is VienNB Mus. 15495, an Alamire manuscript produced sometime between 1508 and 1511 for Emperor Maximilian I and his wife, Bianca Maria Sforza. If this mass was indeed composed for Marguerite of Austria, then it may have been written after she became regent of the Netherlands in 1507 at age 27. The lonely and dolorous sentiments of the chanson *Tout a par moy* make it unlikely that *Missa Faisant regrets* would have commemorated either her brief marriage in 1497 or that in 1501 to Philibert II, duke of Savoy. The death of Philibert in September 1504 seems both the earliest plausible occasion for the mass and also the kind of event for which the chanson text was ideally suited: "I remain quite apart so that I shall not be seen, I shall not appear again, so mournful am I, I keep to myself like a numbed soul." Josquin was by then near the Habsburg-Burgundian court; he had left Ferrara and arrived in Condé only months before, in May 1504. And if it were not enough to be twice widowed by the age of twenty-four, Marguerite then lost her brother, "Philippus Rex Castiliae," in 1506. Thus the years between 1504 and 1507 are particularly likely. These years would also allow for Josquin to have derived the idea of the "fa-re-mi-re" ostinato from Petrucci's 1504 publication of Agricola's setting of Frye's *Tout a par moy* in Canti C. Remembering that Philip's copy of the *Hercules* mass was prepared in 1505, we can plausibly imagine Marguerite, or someone on her behalf, asking Josquin to compose a similar work for her.

In one sense the relationship of Ercole's *soggetto cavato* to the Frye motive might be adduced as evidence for an early dating for the *Hercules* mass. Since Frye's *Missa Nobilis et pulchra* was copied into BrusBR 5557 about 1468, a date in the 1480s may seem appropriate on chronological grounds. There is also a strong organizational parallel between the *Hercules* mass and Busnoys's motet *In hydraulis*, which can be dated between 1465 and 1467. In that motet the contrived tenor also has repeated statements at the fifth and octave: D-C-D, A-G-A, and d-c-d. But these points are not particularly persuasive. The date of the Frye mass is irrelevant given the unique opportunity to relate Ercole's name to a Marian motive, and motivic scaffoldings such as that in the Busnoys motet and the *Hercules* mass occur well into the sixteenth century. Even in the 1460s this kind of scheme must have evoked a consciously archaic aura, harkening back to isorhythmic plans of an earlier generations.

Put the other way around, had Josquin written the mass in the 1480s or 1490s, the exchange of letters in 1502 between Ercole and his two advisers regarding the relative merits of Josquin and Isaac would likely have been different. For if Ercole already had his hexachord mass, surely the envoy in favor of Josquin would have referred to it. This competition between Ercole's advisers over which composer the duke should hire has made it possible to associate a second hexachordal-Marian work with Ercole. In early September 1502 the advocate for Isaac described Isaac's visit to Ferrara, where he wrote a motet in two days on the subject "la-mi-la-so-la-so-la-mi." The kind of composition—a motet on a hexachordal subject—is certainly relevant. That the first syllables would accommodate the name "Maria" is perhaps not coincidental.[33] In fact, Isaac's motet on this motto survives, known in some sources only by the solmization syllables "la mi la sol," and in others, including Petrucci's *Motteti C* of 1504 (1504^1), with a Marian text; Part 1 begins "Rogamus te, piisima virgo Maria," and Part 2, "O Maria, O regina."

Moreover, when Isaac composed a mass on the same motto, the publisher Johann Petreius gave it another Marian association, printing it as *Missa O praeclara* in *Liber quindecim missarum* (1539^1). There the pitches of the eight hexachord syllables are routinely separated into two groups of four, "la-mi-la-sol" and "la-sol-la-mi," and set at times in long note values, at times in short, with an undisguised artificiality comparable to Josquin's techniques in his *Hercules, Faisant regrets,* and *La sol fa re mi* masses. Whether or not Isaac's motivic source was the frottola *La mi la so la so la mi* published in 1509^2, this kind of hexachordal composition is much more typical of the turn of the sixteenth century, when melodic games with text and solmization syllables had become popular in laude and frottole, than it is of the 1480s.[34] The competition between Josquin and Isaac may thus have continued even after Josquin had won the position in Ferrara; his mass for Ercole clearly bests Isaac's. And we know from Glareanus's anecdotal record that Josquin enjoyed this kind of compositional one-upmanship.

These arguments suggest the following chronology:

- In fall 1502 Isaac wrote a hexachordal-Marian motet for Ercole, followed by his mass on the same subject.
- After arriving in spring 1503, Josquin composed his mass for Ercole on a motive that in one context spells Ercole's name and in another—a mass for the duke of Burgundy by Frye—sets the text "ex Maria virgine"; a version of this work found its way to Milan.
- In June 1505 Petrucci published the *Hercules* mass, presumably having acquired it during the previous year.
- In 1505, within a year of Josquin's return north, the court scriptorium copied *Missa Philippus rex castiliae.*
- in 1506–7, following the deaths of Marguerite's husband Philibert II and brother Philip, Josquin wrote his mass for Marguerite on a Frye motive that in one context spells her name and in another sets the text "Faisant regrets," making it one of many *Regrets* compositions associated with her court.
- In the years after 1515 Willaert and Lupus Hellinck wrote their honorific Ferrarese masses and, between 1518–20, whoever planned JenaU 3 brought both *Missa Hercules* (renamed for Frederick the Wise) and *Missa Faisant regrets* (renamed for Frederick's mother) together.

According to this scenario, as opposed to one in which Josquin composed his mass for Ercole in the 1480s, the international popularity of *Missa Hercules dux ferrariae* began to grow almost immediately.

After evaluating all of the traditional kinds of evidence for the date of the *Hercules* mass, the discovery of the Frye citation provides a new approach: the evidence of allusion. Two additional motivic allusions support the 1503 dating. To begin his Kyrie, Josquin apparently modeled the first entrances on Kyrie I of Antoine Brumel's hexachordal *Missa Ut re mi fa sol la* (see Figure 4.11), as suggested by the following: (1) in mm. 1–4, the two upper voices begin, one with the cantus firmus in breves, the other with a contrapuntal idea that is then exactly repeated beneath the next two breves; even the latter half of this counterpoint is similar to Josquin's (the descent by step from a′); (2) in mm. 5–8 of both masses, the superius, contra, and bassus sing without the tenor; and (3) in both masses the tenor enters in m. 9 as the lowest voice. Petrucci printed Brumel's mass as the third mass in *Brumel Misse* (RISM B4643). Josquin would have been interested in this publication

Fig. 4.11 (a) Antoine Brumel, *Missa Ut re mi fa sol la,* Kyrie I; (b) Josquin, *Missa Hercules dux ferrariae,* Kyrie I.

if only because the second mass of the collection was based on his chanson *Berzerette Savoyenne.* Suggestively, the print appeared during Josquin's first months in Ferrara, on June 17, 1503.

The second allusion involves Josquin's six-voice Agnus III, which evidently cites Agricola's most widely circulated work, the song-motet *Si dedero.*[35] Figures 4.12a and b compare the contratenor of the chanson to the beginning and end of the first contratenor statement in the Agnus. Ercole almost certainly knew *Si dedero* shortly after it was composed, since it was copied into the Casanatense manuscript (RomeC 2856) ca. 1480. But while Agricola's text—"I will give sleep to mine eyes, and slumber to my eyelids"—fits well in the third invocation of the Agnus, with its plea "Dona nobis pacem," it seems more appropriate biographically for Ercole in 1503, widowed, in his seventies, clearly at the end of his long life. This later date would also place Josquin's elaboration of the *Si dedero* idea closer to the *Si dedero* masses of Obrecht, Divitis, and Alessandro Coppini, each of which apparently dates from about the turn of the century. Divitis and Coppini may have written theirs while they were colleagues of Agricola: Divitis served with

Fig. 4.12 (a) Alexander Agricola, *Si dedero;* (b) Josquin, *Missa Hercules dux ferrariae,* Agnus III; (c) Jacob Obrecht, *Missa Si dedero,* Agnus III; (d) Alessandro Coppini, *Missa Si dedero,* Et in terra.

Agricola in the chapel of Philip the Fair in 1505–6, and Coppini in Florence in the early 1490s.[36] In their treatment of Agricola's motive, both Obrecht and Coppini shortened the duration of the initial pitch, as did Josquin (see Figures 4.12c and d). However, compared to the chanson and the other mass settings, Josquin's counterpoint is closest to that of Obrecht, especially for Obrecht's Agnus III. Obrecht's mass survives first in MunBS 3154, copied on paper that has been dated 1502.[37] Thus for his six-voice Agnus III, Josquin apparently drew on masses by Obrecht and Brumel that were among the most modern known to him.

In this concluding section of the mass, the complex interplay of texts creates a devout portrait of Ercole: the sung mass text voices the liturgical prayer for mercy and peace, the textual allusions of the ostinato (both sacred and secular) identify the supplicant as Ercole and his willing intercessor as the Virgin Mary, and the contrapuntal chanson allusion conveys a message of rest that might well be interpreted as Ercole's wish for an end to his worldly life.

Considered in the context of Ercole's religious beliefs, the proposed Marian allusion of the cantus firmus makes most sense after the mid-1490s. Ercole's treatment of Savonarola, his considerable support of Suor Lucia di Narni, and his attention to religious functions in his court and the energy with which he built churches in his city all suggest that the intensity of his piety deepened in his last decade. It is perhaps not coincidental that he gave particular expression to Marian beliefs in the years following the death of his wife in 1493 and, not long afterward, the deaths of his daughter Beatrice and daughter-in-law Anna Sforza. These are the years in which he published his *Corona Beatae Mariae Virginis* and established a special Saturday office for the Virgin Mary. As *Nobilis et pulchra* symbolized a mystical wedding between Christ and St. Catherine to honor a Burgundian duke and his bride, it is reasonable to interpret Josquin's achievement in *Missa Hercules dux ferrariae* as a celebration of the overt yet mystical bond between an Italian duke and his beloved Mary.

Notes

I am grateful to Lewis Lockwood and Bonnie Blackburn for their comments on an early draft of this article, and to Joshua Rifkin for sharing with me a prepublication draft of his thoughts on Josquin's mass.

1. Christopher A. Reynolds, "The Counterpoint of Allusion in Fifteenth-Century Masses," *JAMS* 45 (1992): 228–60; and Reynolds, *Papal Patronage and the Music of St. Peter's, 1380–1513* (Berkeley, Los Angeles, and London: University of California Press, 1995), 249–97.
2. Howard Mayer Brown, "Emulation, Competition, and Homage: Imitation and Theories of Imitation in the Renaissance," *JAMS* 35 (1982): 1–48, at 15.
3. Quoted in Michael Baxandall, *Giotto and the Orators: Humanist Observers of Painting in Italy and the Discovery of Pictorial Composition, 1350–1450* (Oxford: Clarendon, 1971), 34.
4. Christopher Reynolds, *Motives for Allusion: Context and Content in 19th-Century Music* (Cambridge, Mass.: Harvard University Press, 2003).
5. Johan Huizinga, *Homo Ludens: A Study of the Play-Element in Culture* (Boston: Beacon, 1950), 4.
6. Ibid.
7. See Reynolds, *Papal Patronage,* 253–57, for reference to several others.
8. Robert Stevenson, "Josquin in the Music of Spain and Portugal," in *JosCon,* 217–46, at 242, names similar masses by Philippe Rogier and Manuel Cardoso as representatives of a Spanish tradition that extends to 1631.
9. I define *motive* as any succession of pitches that is made distinct in one of various ways: by association with a text, by placement at the beginning of a musical phrase (either the beginning of a composition or section of a composition, of a line of text, or even of an interior succession of pitches preceded by rests) or by repetition (either imitatively or in more than one section of a mass, whether as a motto or a cantus firmus).
10. Richard Sherr, "*Illibata Dei Virgo Nutrix* and Josquin's Roman Style," *JAMS* 41 (1988): 434–64, at 445, discusses this coronation motet. The sequence "re-ut-re-ut-re-fa-mi-re" also occurs in the tenor of Frye's *Missa Summe Trinitati* in BrusBR 5557, but in the middle of a melodic unit. Because the tenor of each mass movement and the motet *Salve virgo* begin with an ascending fifth, "G-d-d-d-c-d-d-c-d-f-e-d-c-d," I do not consider the *Hercules* sequence a distinct motive.
11. In *Missa La sol fa re mi* there are occasional deviations from the tenor. On this mass see James Haar, "Some Remarks on the 'Missa La sol fa re mi,'" in *JosCon,* 564–88.
12. Josquin cited Credo I in several masses, including *Missa La sol fa re mi,* where, however, it is not imitated.
13. Willem Elders has commented on this relationship; see his "New Light on the Dating of Josquin's *Hercules* Mass," *TVNM* 48 (1998): 112–49, at 143.
14. Ibid.
15. Lewis Lockwood, *Music in Renaissance Ferrara, 1400–1505: The Creation of a Musical Center in the Fifteenth Century,* Studies in the History of Music 2 (Oxford: Clarendon; New York: Oxford University Press; Cambridge, Mass.: Harvard University Press, 1984), 244.
16. See also the imitative presentations in the Et in terra at "propter magnam gloriam tuam" and in the Patrem at "per quem omnia [facta sunt]."
17. This passage is taken from *Proportionale musices,* book III, chapter 4: "Contratenor autem raro vel numquam primaria pars est. Si tamen super quemvis cantum praecompositum aliquid operari voluerimus primariam efficiemus." The translation is taken from Johannes Tinctoris, *Proportions in Music,* trans. Albert Seay (Colorado Springs: Colorado College Music Press, 1979), 39. I am grateful to Bonnie Blackburn for calling my attention to this passage.

18. Edward E. Lowinsky, "Ascanio's Sforza's Life: A Key to Josquin's Biography and an Aid to the Chronology of His Works," in *JosCon*, 31–75, at 66, n. 112. The observation in Gustave Reese, *Music in the Renaissance* (New York: W. W. Norton, 1954), 244, that Josquin "uses no material peculiar to Agricola" must be qualified to allow for the technique of ostinato on this particular motive.
19. Giovanni Sabadino degli Arienti, *Art and Life at the Court of Ercole I d'Este: The "De triumphis religionis" of Giovanni Sabadino degli Arienti*, ed. Werner L. Gundersheimer, Travaux d'humanisme et Renaissance 127 (Geneva: Librairie Droz, 1972), 35.
20. Lockwood, *Music in Renaissance Ferrara*, 199.
21. Thomas Tuohy, *Herculean Ferrara: Ercole d'Este, 1471–1505, and the Invention of a Ducal Capital* (Cambridge and New York: Cambridge University Press, 1996), 166. See also Edmund Gardner, *Dukes and Poets in Ferrara: A Study in the Poetry, Religion and Politics of the Fifteenth and Early Sixteenth Centuries* (New York: A. Constable, 1904), 456–57.
22. While these are not the only two honorific compositions to receive new dedicatees in the Renaissance, the question of why specific individuals were honored with specific secondhand compositions needs more study. The possibility that the suitability of the motivic allusions played a role in determining both the choice of the original and secondary dedicatees needs consideration. For Agricola's motet *Transit Anna timor*, originally written for Louis XII's wife Anne, the queen of France, the words "Ludovice" and "Gallica" became "Ferdinande" and "Hungarica" when it was copied into VatP 1976–79. However the main reference of the text remained unaltered. Since the new dedicatee was also named Anne (Anne of Hungary), only the references to her husband and country needed revision.
23. On Ercole's stature in Italy, see Tuohy, *Herculean Ferrara*, 19–20.
24. "Te septem propter deposco gaudia virgo/Princeps exhilarans ut Fridericus orat"; see Robert Bruck, *Friedrich der Weise als Förderer der Kunst* (Strasbourg: J. H. E. Heitz, 1903; reprint ed., Nendeln, Liechtenstein: Kraus, 1979), 155. Regarding Frederick's Marian art projects and religious patronage, see also ibid., 78–81, 99–100, 122–25, and 154–56; and Paul Kirn, *Friedrich der Weise und die Kirche: seine Kirchenpolitik vor und nach Luthers Hervortreten im Jahre 1517*, Beiträge zur Kulturgeschichte des Mittelalters und der Renaissance 30 (Leipzig and Berlin: Teubner, 1926), 22–25.
25. Herbert Kellman, "Josquin and the Courts of the Netherlands and France: The Evidence of the Sources," in *JosCon*, 181–216, at 199, n. 64.
26. Lewis Lockwood, "Josquin at Ferrara: New Documents and Letters," in *JosCon*, 103–37, at 108. Lockwood, *Music in Renaissance Ferrara*, 241, later argued that "neither the musical sources nor the circumstantial evidence is really conclusive for any single date." For separate reasons Willem Elders also favored this early period, arguing that it is from 1480; see his "New Light on the Dating of Josquin's *Hercules* Mass," 142.
27. Paul A. Merkley and Lora L. M. Merkley, *Music and Patronage in the Sforza Court*, Studi sulla storia della musica in Lombardia/Collana di testi musicologici 3 (Turnhout: Brepols, 1999), 470. See also Lewis Lockwood, "Introduction," in *The Collected Works of Josquin des Prez*, vol. 11: *Masses Based on Solmisation Themes*, xi–xii (Utrecht: Koninklijke Vereniging voor Nederlandse Muziekgeschiedenis, 2002; I am grateful to him for sharing a typescript of his remarks before publication); Edward E. Lowinsky, "Josquin des Prez and Ascanio Sforza," in Maria Luisa Gatti Perer, ed., *Il duomo di Milano: Congresso internazionale, Milano, Museo della scienza e della tecnica, 8, 12 settembre 1968, Atti*, Monografie di arte lombarda/I monumenti 3 (Milan: La Rete, 1969), 2:17–22, esp. 19; and also Lowinsky, "Ascanio Sforza's Life," 59–60. On stylistic grounds Patrick Macey suggests "that the work was written as early as the 1480s"; see his "Some Thoughts on Josquin's *Illibata dei virgo nutrix* and Galeazzo Maria Sforza," in Albert Clement and Eric Jas, eds., *From Ciconia to Sweelinck: Donum Natalicium Willem Elders*, Chloe/Beihefte zum Daphnis 21 (Amsterdam and Atlanta: Rodopi, 1994), 111–24, at 122.
28. Helmut Osthoff, *Josquin Desprez* (Tutzing: Hans Schneider, 1962–65), 1:177. Arguments for considering this mass a later work have been advanced by Joshua Rifkin (unpublished study); Jeremy Noble, "Josquin Desprez," *New Grove*, 9:723; and Lockwood, "Introduction," xii.
29. Merkley and Merkley, *Music and Patronage in the Sforza Court*, 330.
30. Knud Jeppesen, "Die 3 Gafurius-Kodizes der Fabbrica del Duomo, Milano," *Acta Musicologica* 3 (1931): 14–28.
31. See Reynolds, *Papal Patronage*, 104–9, and 103 n. 60 for reference to other manuscripts.

32. Rifkin's unpublished findings are discussed in Elders, "New Light on the Dating of Josquin's *Hercules* Mass," 116, and David Fallows, "Josquin and Milan," *Plainsong and Medieval Music* 5 (1996), 69–80, at 75 n. 19.
33. Such an interpretation of the "la-mi-la" cantus firmus of Josquin's *Illibata dei virgo* has been suggested; see Lowinsky, "Ascanio Sforza's Life," 59; and Macey, "Some Thoughts on Josquin's *Illibata dei virgo nutrix.*"
34. See Knud Jeppesen's comments in Jeppesen and Viggo Brondal, *Die Mehrstimmige italienische Laude um 1500* (Leipzig: Breitkopf & Härtel, 1935), lii–liii. On Isaac's mass and motet, see Martin Staehelin, *Die Messen Heinrich Isaacs* (Bern and Stuttgart: Paul Haupt, 1977), 3:63–67; and Heinrich Isaac, *Messen,* ed. Martin Staehelin, Musikalische Denkmäler 7–8 (Mainz: B. Schott, 1970–73), 2:171–72.
35. On this work, see Honey Meconi, "Sacred Tricinia and Basevi 2439," *I Tatti Studies* 4 (1991): 151–99.
36. Frank D'Accone, ed., *Music of the Florentine Renaissance,* CMM 32 (Rome: AIM, 1966–96), 2:xii, proposes the turn of the century for the date of Coppini's mass based on the presumed date of its source, Librone 3 (MilD 3). But if Coppini composed his mass in Florence in the early 1490s, then his five-voice mass probably antedates the four-voice masses of Obrecht and Divitis.
37. Regarding Obrecht, see Thomas Noblitt, "Die Datierung der Handschrift Mus. ms. 3154 der Staatsbibliothek München," *Musikforschung* 27 (1974): 36–56.

5
Habsburg-Burgundian Manuscripts, Borrowed Material, and the Practice of Naming

HONEY MECONI

Nomina sunt consequentia rerum.
Dante, *La vita nuova*[1]

While preparing the entry on Pierre de la Rue for the revised *New Grove* I was struck by the multiplicity of names for his masses.[2] Which was the "right" name, the one under which the mass should be listed in the works list and used for discussion? For the most part I ended up following traditional scholarly practice—that is, using the name of the borrowed material—but started wondering about both the justification for that practice and the nature of names in the fifteenth and sixteenth centuries. With motets and secular works typically taking names from their texts, questions about naming must focus primarily on masses. In a separate essay I explored Petrucci's role in the naming of things;[3] here the case study is of the very large and very significant collection of Habsburg-Burgundian court manuscripts. The masses of La Rue, which dominate these sources, make an excellent starting point.

La Rue's masses circulated predominantly (about a third of them exclusively) in manuscripts prepared by the scriptorium associated with the court of Habsburg-Burgundy, his employer for the final twenty-four or more years of his working life. His masses thus have a pedigree of authenticity unrivaled among his major contemporaries. These Habsburg-Burgundian manuscripts (hereafter "court sources" or "court manuscripts"), prepared as gifts or on commission, are widely known as among the most elaborate and beautiful sources for polyphonic music of the time. In addition to calligraphy of the highest order, amusing drolleries, painted miniatures, and the frequent appearance of elaborate borders, the visual appeal of these collections was enhanced by extensive use of rubrification, most often in voice parts to highlight the use of preexistent material. It is from these indications that many masses assume the names we use today.

The practice of indicating preexistent material this way neither started with nor was limited to these court manuscripts (though not all scribes elsewhere used red ink for this purpose), but these sources were exceptionally generous in providing this extra material and rubricating it. Most often these indications were confined to the initial opening of a mass, but they sometimes extended through an entire composition. Needless to say, this auxiliary matter can provide extremely helpful clues for the analysis of a work.

What becomes evident when looking at these rubrics and at the titles of masses, however, is that the formal name of the mass as indicated in the title (or alternatively, in the index) was by no means always the same as the name of the preexistent material, whether that was given in the voice parts or not. This is abundantly clear for the works of La Rue in court sources. One might reasonably assume that these manuscripts, many dating from his lifetime and possibly prepared under his supervision, present his compositions with their proper names. It comes as a surprise, then, to see the considerable variety of identities these works had, and the ways in which they differ from our modern namings.

Some of La Rue's masses are titled in court sources as modern scholars usually do, with the name of the primary preexistent material. Most often this is prefaced with the designation "Missa" (such as *Missa Cum iocunditate* in 's-HerAB 72B or *Missa Tous les regretz* in SubA 248). Sometimes "super" or "supra" are added (e.g., *Missa super Inviolata* in JenaU 7);[4] the number of voices might appear as well (*Missa quinque vocum supra incessament* in JenaU 4). At times the model name is used by itself as the title (e.g., *Lomme arme* in VienNB 1783).

Other masses appear with "Missa" followed only by the number of voices, such as *Missa quinque vocum* for the appearance of *Missa Alleluia* in MontsM 773. This kind of title is used not just for masses exceeding the usual component of four voices, but even for those as well; thus *Missa quatuor vocum* is the name for *Missa O gloriosa domina* in JenaU 5.

A third category of mass title provides an indication of the devotional function of the mass (e.g., *Missa pro fidelibus defunctis* in VatS 36). Sometimes this included the number of voices as well (e.g., *Missa pascale quinque vocum* in BrusBR 15075), occasionally with the initial emphasis on the performing forces (e.g., *Missa quinque vocum de septem doloribus beatissime marie virginis* in BrusBR 215-16, where the title distinguishes it from an anonymous Seven Sorrows mass for four voices in the same manuscript). La Rue was unusually interested in composing masses with an obvious liturgical connection, so it is no surprise to find this a common choice for titles. Further, it is an unambiguous one for a work using multiple models, as is the case with all three of the pieces just cited. In several cases this devotional purpose trumps the preexistent material when it comes to titling the mass. Thus, the mass in VatS 36 whose tenor includes a text beginning "O quam pulchra" is titled *Missa de virginibus,* while the work that makes extensive use of the text and melody of *Nos autem gloriari* is always identified in the manuscripts as *Missa de sancta cruce.* In another instance the tenor rubric "Felix Anna" in VienNB Mus. 15496 gives way to the title *Missa de Sancta Anna* in that manuscript.

Occasionally a mass will have a title that is obscure (at least to us), as with *Missa Almana,* identified thus in VatC 234 and VienNB 1783, but the most common presentation for La Rue's masses in court sources was without any title whatsoever. This is true even after discounting the many masses lacking their beginning folio in collections mutilated for their beautiful miniatures. Sometimes an elaborate decorative border overran the space where a title would normally appear, but titles could also manifest themselves between the superius and tenor voices (as with *Missa Almana* in VienNB 1783) or sometimes within the tenor voice itself (the same work in VatC 234); a border did not have to prevent a scribe from adding a title.[5] Thus, a formal title wasn't seen as a prerequisite for the presentation of a mass, and six of La Rue's works (the *Missae Conceptio tua, de Sancto Antonio, Sancta dei genitrix, Sine nomine I, Sub tuum presidium,* and *Tandernaken*) lack titles in any court manuscript; model names or auxiliary material—if present at all—appear within the voice parts only (a seventh mass, *Missa O salutaris hostia,* lacks a title in its sole court source, MontsM 773, but the portion of the manuscript where any title might have appeared is now missing).

Individual masses could also exist under different titles in different court manuscripts. For example, the Subiaco manuscript transformed two works known elsewhere by their preexistent material into masses identified by devotional role; thus *Missa super Ave Maria* in VienNB Mus. 15496 is *Missa de annuntiatione maria* in SubA 248 (it is without title in MechAS s.s.), while *Puer natus est* (so called in the JenaU 22 index, without "Missa") is the *Missa de nativitate* in SubA 248 (and lacks a title in VienNB 1783). In what would seem an unnecessary change, *De beata virgine* (again the index indication in JenaU 22, and also without title in VienNB 1783) parades as *Missa de domina* in SubA 248. As for the mass we currently call *Missa Ista est speciosa,* no court manuscript identifies it as such; JenaU 2 titles it *Missa de sanctissima virgine Maria* and two simply call it *Missa quinque vocum* (JenaU 4 and BrusBR 15075; three other sources are missing the section normally used for titles, while the Pleni also appears as a bicinium—without a corresponding mass title—in VienNB Mus. 18832).

While one might posit that La Rue, as quasi-official court composer, received some sort of special treatment as far as titles went in court manuscripts, the works of other composers therein are handled similarly. Thus, for example, Hottinet Barra's *Missa Ecce panis angelorum* appears in BrusBR IV.922 as *Missa de venerabili sacramento,* the same title used for Josquin's *Missa pange lingua* not when it appears in that manuscript, as one might expect (the collection is largely devoted to works venerating the sacred host), but rather in VienNB 4809. Barbireau's *Missa Virgo parens Christi* likewise becomes *Missa de venerabili sacramento* in MontsM 766. Bauldeweyn's *Missa Da pacem domine* is merely *Missa quattuor vocum* in MunBS 7. Josquin's *Missa de beata virgine* is *Missa de domina* in VienNB 4809. Numerous masses lack any title whatsoever, though usually, as with La Rue's masses, the model name shows up within the tenor or other voice parts. In the sixteenth-century index added later to VatC 234, while Ockeghem's masses are identified therein by name, those by other composers are simply "Missa de" followed by the name of the composer, even when the composer has more than one work included. The de Orto work that Petrucci calls *Petita camuseta* shows up in VienNB 1783 as *Officium My my.* The mass sections that appear in the bicinia collection VienNB Mus. 18832 and the chansonnier FlorC 2439 are typically not identified as coming from masses at all. And so on.

Similar titular alterations take place in sources outside the court complex. La Rue's *Missa Almana* goes by *Missa pourquoy non* in the index of VatS 45 (referring to the presumed—albeit nonexistent—relationship between that mass and La Rue's famous chanson of the same name), while in Petrucci's 1503 collection devoted to the composer (RISM L718) the same mass is *Sexti ut fa,* from its tonal orientation and the hexachordal designation of the opening gesture. His *Missa de beata virgine* becomes *Missa super coronatum* in DresSL Pirna IV, with the Credo transformed into *Patrem Cardinal* in BolSP 38; *Missa Sub tuum presidium* is *Quarti toni* in 1515[1]. Extending our survey again beyond La Rue, we see multiple means of identification in noncourt manuscripts. To cite a few of the possibilities: some have the name of the borrowed material as the title (e.g., *A l'umbretta d'um bussonet* by Brumel in TurBN I.27), some are known just by the name of the composer (e.g., *Messa de rinaldo* in BolC Q19), some are modal (e.g. *Missa Gaffori primi toni brevis* in MilD 2, where the genre is also called forth), some draw attention to the fact that they are complete masses (e.g., Isaac's four-voice *Missa Wohlauf Gesell von hinnen*, which is presented in MilD 3 with neither title nor identification of the model but named in the index as "Kyrie cum tota missa de Enrich ysaac"), and many have no title at all (e.g., the untitled work in SienBC K.I.2 that we call *Missa De tous biens plaine*).

This short overview demonstrates the presence of a multiplicity of possible identities for a fifteenth- or sixteenth-century mass, most especially one based on sacred material.[6] The contemporary inconsistency about the names of masses (or even the necessity for them) suggests not so much confusion about their identity but rather flexibility, a flexibility equally apparent in scribal treatment of underlay, ficta, and occasionally the very makeup of the mass itself.[7] Our modern preference for names giving a mass's model obscures the fact that it was far from universal practice in the fifteenth and early sixteenth centuries.

While scribes only sometimes elevated the name of the preexistent material to the status of a formal title, they were far from averse to the idea of identifying that material. In fact, contemporary manuscripts show just how important that idea could be. On more than one occasion scribes blazoned forth the name of the wrong model in the voice parts of a mass. La Rue's *Missa Cum iocunditate*, which appears without title in MilA 46 (not a court manuscript), gives the word "Dirige" in each voice at the opening of the mass in this manuscript. The scribe, unaware of the origin of the work's ostinato and the mass's function (for the Nativity of the Virgin), evidently thought that the model was the antiphon *Dirige domine deus meus* from the Office for the Dead, since its melodic opening is the same as that of the antiphon *Cum iocunditate*. Similarly, Févin's mass given in MunBS 7 as *Missa de beata virgine Salve sancta parens* holds a name that is probably misleading; Edward Clinkscale found no resemblance between the mass and the *Salve sancta parens* chant (the beginning of the Introit for general Marian masses) beyond the first few notes.[8] In an instance where the scribe should certainly have known better, La Rue's *Missa de beata virgine*, given without title in VienNB 1783 (a court manuscript compiled while La Rue was employed there), has the rubric "Salve sancta parens." While that chant indeed begins like Kyrie IX, the model for the first movement of this mass, it soon diverges, and thus the rubric is ultimately misleading.

Mouton's mass often known as *Missa Regina mearum* provides a more complicated example of confusion. This popular work circulated in a dozen surviving sources; all but two (including printed collections by Attaingnant and Susato) designate it as *Missa de almania* or some variant thereof. The dissenting voices are Petrucci's 1515 collection of Mouton masses (RISM M4015), where it acquires the title *Regina mearum*, and StuttL 46 (same title; perhaps derived from Petrucci). Andrew Minor, editor of the collected-works edition, said that neither the German mass title nor *Regina mearum* could be explained in connection with the mass, and suggested rather that the words appearing in the Kyrie of the work in CoimU 2, "Adieu solas, adieu joye," referred to a song that was the real basis for the mass. Since he could not find a chanson beginning with that text, however, I wonder whether these words are yet another scribal red herring, evidence of the need to designate some—any—model for a work. Certainly our own use of the *Regina mearum* title is based on either an unjustified faith in Petrucci or an apparently mistaken belief that it is the name of the model of the mass.

In their haste to proclaim the model of a work, then, some scribes made errors that should put modern scholars on their guard about these presumed models. In contrast to such carelessness, however, scribes of the court manuscripts could be painstaking about how they represented borrowed material within the voice parts of a mass. In some instances a special, separate scribe copied the cantus firmus voice, dividing notes as necessary to match the model text.[9] In others the scribe altered the rubric for the borrowed material to take into account some aspect of the manuscript's appearance, as for example in the treatment of Pipelare's *Missa Fors seulement* in MechAS s.s. Here the scribe has conflated the first two lines of the cantus firmus text, with the somber "I die" replaced by the more positive "I remain," so as not to clash with the miniature on the same opening of the (then very much still living) archduke Charles.[10]

As interesting as these changes are, a more significant practice is the scribal habit of indicating the model in more than one voice at the opening of the mass, in contrast to the more usual procedure of giving the model name in merely a single voice. Of the thirty-eight complete court manuscripts that contain masses, only five eschew this practice: the early collections VatC 234, JenaU 22, and VerBC 756, and the later volumes BrusBR 215–16 and VienNB Mus. 15496 (two other manuscripts, JenaU 9 and VatS 34, are missing the parts that would have contained the proper indications). All other manuscripts mark multiple borrowings for one or more masses of their collection. And rather than being just an example of scribal overkill, the practice often (though not always) demonstrates a telling knowledge on the part of the scribes (or whoever copied the exemplars that they used) about the material with which they were working.

About half the time when rubrics appear in multiple voices on the initial opening of a mass the scribe has gotten it exactly right.[11] In the openings of at least thirty-four masses the scribe has marked each voice in which borrowed material appears with the name of that borrowed music. One such example is seen in BrusBR 15075 with La Rue's *Missa Ave sanctissima Maria,* a six-voice canonic parody mass based on his own six-voice canonic motet of the same name. Here the mass is simply titled *Missa sex vocum,* while each voice part includes the model name squeezed in among the Kyrie text; the recto of this opening is shown on the cover of this book.

Less frequently the name of the borrowed material appears in superius and tenor voices only, even though the remaining parts also employ the model. At least thirteen masses do this, including Févin's *Missa Mente tota* (based on Josquin's motet) in JenaU 3 and Gascongne's *Missa Mijn hert* (based on La Rue's song) in MunBS 7. Pieces such as these are different from works such as Agricola's *Missa Pascale* in VienNB 1783 or Moulu's *Missa Alma redemptoris mater* in 's-HerAB 72B, where the model material appears only in superius and tenor and thus warrants texting only there. In "undertexted" masses such as Févin's and Gascongne's, all voices borrow from the model but only superius and tenor are so marked. The practice is likely a reflection of the generally fancier decoration traditionally given to the verso side of a mass's opening, with rubrification adding one more element to tempt the eye.

A different type of multiple "undertexting" shows up in at least five masses, as for example in the case of the anonymous *Missa O werde mont* in MunBS F. Here only altus and bassus have the "O werde mont" text despite the fact that superius and tenor as well commence like the model, in this case the four-voice song that appears in LonBL 31922. Other examples have similar discrepancies. In rare cases scribes were overenthusiastic about proclaiming the name of the model, such as in Bauldeweyn's mass in 's-HerAB 72B, titled there *Missa sex vocum quam pulcra es*. Here only four of the six voices are drawn from Bauldeweyn's own four-voice motet of the same name, but all six mass lines are texted with the model's name on the initial opening.

Just as the title of an individual mass could vary from manuscript to manuscript in the court sources, so too did the frequency with which a model name was written by the scribe. La Rue's five-voice *Missa Conceptio tua* provides a good example of this variety. Each of the three manuscripts that contain the opening of the mass (four other sources have a mutilated or missing opening) provides different indications for the borrowed material. In MechAS s.s., four of the five voices—the four that present the chant model in imitation—give the name of the chant, a precise fit between model use and model indication. JenaU 4 goes overboard, with all voices—one too many—marked with the model name, while BrusBr 15075 indicates the model only in the first tenor, the part that throughout the mass will be most devoted to the chant. Or take the case of La Rue's *Missa Ave Maria*, also chant-based. In MechAS s.s. and in SubA 248 all voices are marked at the start with the name of the model, which appears in imitation in each part. In VienNB Mus. 15496 only the tenor is marked with the model, while in MontsM 773 no voices are marked. In JenaU 12 the verso page (containing the superius and tenor) is missing, leaving for the first opening only the altus (marked with the model name) and the bassus (no model indication, even though the model is present).

The primary purpose behind multiple indications of the model does not seem to be naming.[12] Not only would a single indication have sufficed (in any voice, for that matter),[13] but some masses practicing this multiple texting already had titles, in certain cases the name of the model itself (as in Bauldeweyn's mass on *Quam pulcra es,* indicated above, or La Rue's mass titled *Missa Incessament* in 's-HerAM 72B, with "Incessament" given in each voice as well). Instead, the highlighting of the model seems to have functioned both as an extra layer of visual luster as well as—and, I would suggest, more significantly as—a means of calling attention to the presence of borrowed material, usually with remarkable accuracy.[14] The type of model was not important, but rather the fact that its material showed up at the commencement of the work in more than one voice. Thus, in JenaU 21, Josquin's work titled *Missa de venerabili sacramento* has "Pange lingua" inscribed at the beginning of each voice; the mass begins with the chant (monophonic model) in imitation in each voice. In MontsM 766 the anonymous *Missa memor esto quinque vocum* has the words "Memor esto" in each voice along with the imitative motive from the beginning of Josquin's motet. In VienNB 11778, each voice of Josquin's mass on *Fortuna desperata* opens with material from the (nonimitative, homorhythmic) song that is its model, and each voice has the text "Fortuna desperata." It didn't matter to the scribes, then, whether the mass was a cantus firmus, paraphrase, or parody mass, or whether the model was imitative or not. The important thing was that more than one voice of the mass was derived from the model, and as noted above, it was rare for them to get this wrong.

Court scribes were not so obsessed with borrowings that every scrap of old material was signaled in these manuscripts (certainly that would make life much simpler for musicologists). Indications of multiple borrowing usually fail to extend beyond the opening, and when they do, they are typically restricted to a single voice that most probably used the non-Ordinary text of the model in performance. One could argue that the borrowing indicated on a mass's opening folio was the material least in need of attention: it appears at the beginning of the work, it is more likely than any other borrowed material to be presented without paraphrase, and it is sometimes already announced in the title. It seems, then, as if the purpose of these rubrics was to alert rather than to lecture. The performers (and the manuscript's owner, if he or she perused the volume) were duly warned that the piece to ensue was built on something else, and possibly even built polyphonically. It was up to them to catch the later references, which were normally unannounced visually.

I should stress again that scribes were content in many manuscripts to limit indication of the borrowed material to a single voice (if that), even if preexistent music actually showed up in more than one voice at the start. But there are suggestions that over the years there was an increase in scribal concern with "borrowing polyphonically," that is, borrowing in more than one voice, whether the model was polyphonic or monophonic. As noted above, three of the earliest court manuscripts eschew any indications of borrowing polyphonically, and of those that do mark this material, only superius and tenor voices—on the visually more important verso of the initial opening—receive the extra indications.

The tide shifts with the collection VienNB Mus. 15495, compiled between 1508 and 1510 for the emperor Maximilian and his wife, Bianca Maria Sforza. Three of the seven masses here restrict rubrification to the tenor voice, even though two of them (Josquin's *Missa Faisant regrets* and his *Missa Une mousse de Biscaye*) have model material in at least one more voice. A fourth mass, Thérache's *Missa O vos omnes,* correctly puts the model text in the superius, the voice using the preexistent material. But the remaining three masses expand their rubrification beyond that seen in earlier court manuscripts. Bruhier's *Missa Mediatix nostra* has that text in both tenor and bassus, each of which uses music from the chant. Compère's mass on *Alles regrets* gives that text in altus, tenor, and bassus voices (which borrow from the tenor, tenor, and superius, respectively, of Hayne's famous chanson) but curiously neglects the superius voice, which draws extensively on the model's superius. Finally, Févin's *Missa Mente tota,* widely recognized as one of the earliest true parody masses, has its model indicated in all voices. There is clear awareness here of where the model is, with a desire to show it as the mass opens. Court manuscripts from here on pick up on this intensified awareness. This is to say not that all later court manuscripts labeled all borrowing and always did so correctly, but rather that we now find numerous multiple labels clearly linked to multiple borrowing, unlike the earlier, visually oriented marking on the verso of a mass's opening.

We cannot say for certain what prompted this more sophisticated treatment of borrowed material, but two things in particular stand out in connection with VienNB Mus. 15495. The first is that at the time of the manuscript's compilation master scribe Petrus Alamire was finally formally associated with the court, and we may be seeing the result of his influence on manuscript production.[15] The second is that with Févin's mass we have firm evidence of the court's exposure to polyphonic borrowing on the largest scale.

To say that the origins of what is often still called "parody" are controversial is to put it mildly. Nor is the chronology of full-fledged masses of this type at all clear. But the scribal treatment of borrowed material in VienNB Mus. 15495 and the timing of the manuscript's compilation are extremely suggestive. VienNB Mus. 15495 is the earliest firmly dateable source for any mass generally considered a true parody mass. While TurBN I. 27 contains Févin's *Missa Ave Maria,* another very early parody mass, the currently suggested date of ca. 1500 for this collection is based on repertoire rather than codicological elements and thus is not fixed.[16]

It is therefore very possible that court musicians and scribes first encountered a real parody mass with the arrival of Févin's *Missa Mente tota.* This statement requires some explanation. Divitis, after all, composed a parody mass on Agricola's *Si dedero,* and some have suggested that this mass was created when the two were colleagues at court in 1505–6.[17] Given the model's phenomenal popularity, though—it was one of the most beloved works of the time, circulating in more than twenty sources for a good sixty years—Divitis's appropriation of such a standard work hardly required personal contact.

La Rue's own forays into polyphonic borrowing must be accounted for as well, since he wrote a series of increasingly derivative masses that drew on his own models, culminating with the six-voice canonic *Missa Ave sanctissima Maria.* Yet despite the difficulty of providing dates for most of his compositions, various factors suggest that his most extensive use of parody procedure comes from the later part of his court employment, after he returned north from Spain in 1508—just in time for exposure to new compositional currents coming from the French court by way of Févin's masses.[18]

I would suggest, then, that the court's exposure to this new creature, the parody mass, made all musicians and scribes more conscious of borrowing polyphonically, and that the manuscripts reflected this consciousness, with multiple citations of the model no longer restricted in these collections to the more important verso of an opening. Certainly the scriptorium embraced the new genre. The following list of twenty of the earliest parody masses shows that court manuscripts transmitted more than half of these. Here, masses appearing in court manuscripts are marked with asterisks, with a double asterisk signifying masses whose earliest surviving source is (or could be, in the case of dating vagaries) from the court: Févin, **Missa Ave Maria, Missa Dictes moy toutes,* ***Missa Mente tota,* **Missa Sancta trinitas;* Divitis, *Missa Gaude Barbara, Missa Quem dicunt homines,* ***Missa Si dedero;* Richafort, ***Missa O genitrix, Missa Veni sponsa Christi;* Mouton, *Missa Quem dicunt homines;* Bauldeweyn, ***Missa En douleur en tristesse,* ***Missa Quam pulcra es;* Forestier, ***Missa Baises moy,* ***Missa Intemerata virgo;* Gascongne, *Missa Benedictus,* ***Missa Mijn hert, Missa Mon mari ma diffamee, Missa Nigra sum,* **Missa Pourquoy non.*

This list suggests other interesting chronological ideas. Richafort worked at Sint Rombaut's in Mechelen from late 1507 to 1509, after which he was succeeded by Noel Bauldeweyn, who remained there until 1513. During this entire period the Habsburg-Burgundian court used Mechelen as one of its two main bases of operation (Brussels was the other), remaining there for very extended periods of time and using the chapel of Sint Rombaut on their own. One wonders whether Richafort and Bauldeweyn, as well as La Rue, were thus exposed to new currents from France via the music that ended up in the court manuscripts.

* * *

Many of the naming practices observed in the Habsburg-Burgundian court manuscripts appear in other manuscripts of the time as well, although only the same kind of detailed investigation undertaken here can demonstrate how widespread these practices are. But it should be clear that, even though we are seeing everything through the scrim of a scribe's practice, there is no "right" name for a mass from this time, that multiple identities were possible, that even the need for a name was not universal in contemporary manuscripts. The same is true as well for the naming of borrowed material within a mass's voice parts. It was often important enough for the scribe to take considerable pains with its presentation, but it was not practiced uniformly in all manuscripts, or even consistently within a single collection.

Why, then, do modern scholars use the name of the borrowed material to identify the mass when possible, even when, as in the case of certain masses, none of the surviving copies used the model's name? Inertia plays its part; once a mass is titled in any kind of definitive volume (e.g., a collected-works edition or an encyclopedia entry) it is hard for later writers to break ranks without generating confusion. This is especially true since most published material, including critical notes to editions and modern manuscript indices, only rarely indicates exactly how a work is denoted in its sources, making it extremely difficult to determine its name without access to microfilms or the sources themselves. By default, then, we are normally forced to refer to already established names.

These names also accord with our long-standing preoccupation with borrowed material as an analytical tool. A model is one of most distinctive ways to identify a piece: calling La Rue's mass *Missa Ista est speciosa* lets us zero in on the cantus prius factus, while calling it *Missa de sanctissima virgine Maria* merely informs us that it is a Marian mass. To ignore borrowing in the analysis of most Renaissance masses would result in an incomplete understanding, neglecting the very element that the composer began with. The study of borrowing remains one of the most direct ways that we can talk about this music.

Nor are we modern scholars alone in our concern with borrowed material. It is, after all, one of the contemporary options for titling an early mass. Scribes' careful indication of model material, especially in more than one voice, shows this interest as well. But so too does the very transmission pattern of fifteenth- and sixteenth-century music. It is rare for a mass without any preexistent material to achieve popularity: *sine nomine* masses and masses with only modal identities circulate most often as *unica.* Masses on known melodies are typically the ones that are copied more than once. As for masses that borrow polyphonically, it is striking how often these works use models that are extremely popular in their own right. To cite just a few examples: *Tous les regrets, Incessament, Pourquoy non,* and *Mijn hert,* secular compositions by La Rue that were models for his own and Gascongne's masses, were all widely circulated works, while Févin based works on Josquin's popular motets *Ave Maria* and *Mente tota,* as well as his own well-known *Sancta trinitatis.* Contemporary listeners apparently responded with interest to models as well.

Finally, the historiography of early music has fostered our emphasis on named masses and borrowed material. Writers on music such as Burney and Hawkins turned to printed sources far more often than manuscripts when seeking information; theirs was a culture used to both printed material and clearly identified compositions. Glarean and Petrucci were of particular importance as sources of material for early historiographers.[19] As I have shown elsewhere, Petrucci initiated the practice of providing a name for every mass that he published,[20] and by the time printing really got under way, in the 1530s, parody was the most common procedure for mass composition, with works almost uniformly named after their models. At the same time, Glarean (and other sixteenth-century theorists) drew heavily on printed sources of music.[21] Thus, the reliance by Burney, Hawkins, and those who followed them on published music and printed theoretical treatises helped establish formal names, especially those based on model material, for sixteenth-century masses. The virtual sanctification in the early nineteenth century of Palestrina—with his preference for parody on clearly identified models—only solidified this tendency. Some mass names were enshrined before any of the relevant manuscript sources were consulted, and masses were known by name long before their music was published—a situation still true for many composers today.

"Names are the consequence of things," Dante tells us. Contemporary names for masses, when they existed at all, could reflect many things, depending on what the individual scribe or printer wished to stress. Only one of these was the work's model. Our modern names for Renaissance masses are a consequence of the historical circumstances of the rediscovery of early music as well as our own preoccupation with models. The multiple naming of borrowed material in a mass's voice parts could reflect a desire for increased visual appeal on a mass's first folio, or a growing interest in identifying the extent to which borrowed material pervaded a new work, sparked by the first generation of true parody masses. We owe it to ourselves to pay much closer attention to how compositions were identified in the sources, whether by means of titles or callouts of borrowed material, for we neglect to do so at the risk of masking the complexities of identity and construction that truly characterize this music.

Notes

1. Dante Alighieri, *Vita Nuova: Italian Text with Facing English Translations*, ed. Dino S. Cervigni and Edward Vasta (Notre Dame and London: University of Notre Dame Press, 1995), 68–69.
2. Honey Meconi, "La Rue, Pierre de," *New Grove II*.
3. See Honey Meconi, "Petrucci's Mass Prints and the Naming of Things," in *Proceedings of the Conference Venezia 1501: Petrucci e la stampa musicale* (Venice: Fondazione Levi, forthcoming), for a discussion of Petrucci's practices and their ramifications. The very act of naming a mass at all was sporadic before Petrucci, the first to make sure all of his masses had some kind of title in his publications.
4. "Super" or "supra" indications were used with both monophonic and polyphonic models.
5. In what appears to be a unique case in the court manuscripts, La Rue's *Missa L'homme armé I* in BrusBr 9126 has the words "Lomme arme" appearing within the superius voice of the opening Kyrie and functioning as a title; in contrast to the other voices in the Kyrie, it does not use the model.
6. Meconi, "Petrucci's Mass Prints," gives examples of secular renamings.
7. This seems to be especially true concerning Easter masses, but it happens in other works as well. In BrusBR IV 922 the Easter mass by Isaac is actually a conflation of two of his *Missae paschale*. Agricola's Easter mass, which has no Credo of its own, appears twice with the Credo from Isaac's *Missa Tmeiskin was jonck* (in JenaU 22 and VerBC 756) and once with his own *Credo Je ne vis oncques II* (VienNB 1783; that Credo appears on its own in VatS 41). This same mass uses a Sanctus in all three sources that is ascribed to Verbonnet in VerBC 756 but nowhere else (this Sanctus is known only through the Agricola mass). The Credo from Isaac's *Missa Tmeiskin* is also used to replace the normal Credo of Brumel's *Missa de virginibus* in VienNB 1783 (presented without title). Josquin's *Missa pange lingua* in BrusBR IV 922 takes its Benedictus from Gascongne's *Missa Es hat ein Sin* and its Pleni from an unidentified piece, while Obrecht's *Missa Sicut spina rosarum* in VatS 160 replaces its own Credo with La Rue's (attributed but untitled) *Credo de villagiis*. Such mixtures are not unknown outside court manuscripts; in BolC Q19, for example, the *Messe de brumel* is made up of the Gloria and Credo of his *Missa Descendi in hortum* and the Sanctus and Agnus of his *Missa dominicalis*.
8. See Edward Henry Clinkscale, "The Complete Works of Antoine de Févin" (Ph.D. dissertation, New York University, 1965), 1:100ff.
9. See Flynn Warmington, "A Survey of Scribal Hands in the Manuscripts," in Herbert Kellman, ed., *The Treasury of Petrus Alamire: Music and Art in Flemish Court Manuscripts 1500–1535* (Ghent and Amsterdam: Ludion, 1999), 41–46, at 41 (2) and 45 n. 7.
10. See Bonnie J. Blackburn, "Messages in Miniature: Pictorial Programme and Theological Implications in the Alamire Choirbooks," *Yearbook of the Alamire Foundation* 5, forthcoming. I am grateful to Professor Blackburn for kindly sharing the relevant passages prior to publication.
11. Precise figures cannot be established, since models for some masses remain unidentified and various manuscripts have lost material through mutilation or other trials.
12. Nor, in most cases, does it seem to have been for performance.
13. See the example given in note 5 above.
14. So much so that I wonder whether explanations will eventually surface for some of the instances now seen as inaccurate.
15. On Alamire, see Eugeen Schreurs, "Petrus Alamire: Music Calligrapher, Musician, Composer, Spy," in Herbert Kellman, ed., *The Treasury of Petrus Alamire: Music and Art in Flemish Court Manuscripts 1500–1535* (Ghent and Amsterdam: Ludion, 1999), 15–27.

16. See Lawrence F. Bernstein, "*La Courone et fleur des chansons a troys:* A Mirror of the French Chanson in Italy in the Years between Ottaviano Petrucci and Antonio Gardano," *JAMS* 26 (1973): 1–68, at 14.
17. See for example Martin Picker, "Divitis, Antonius," *New Grove II*, 403. For revised dates of Divitis's court service, see Honey Meconi, *Pierre de la Rue and Musical Life at the Habsburg-Burgundian Court* (Oxford: Oxford University Press, 2003), 75.
18. On the chronology of La Rue's masses, see Meconi, *Pierre de la Rue*, 97–115.
19. For a demonstration of the importance of printed sources for eighteenth- and nineteenth-century writers on music, see the historiography for La Rue given ibid., 214–34.
20. Meconi, "Petrucci's Mass Prints."
21. The most thorough examination of this phenomenon is found in Cristle Collins Judd, *Reading Renaissance Music Theory: Hearing with the Eyes* (Cambridge: Cambridge University Press, 2000).

6

Aspects of Musical Borrowing in the Polyphonic *Missa de feria* of the Fifteenth and Sixteenth Centuries

ANDREW H. WEAVER

During the more than twenty years since the publication of Howard Mayer Brown's influential article on the musical relationships between selected fifteenth- and sixteenth-century secular pieces, the study of borrowing has taken great strides toward a fuller understanding of both compositional process and musical meaning in the Renaissance.[1] Brown's article laid the groundwork for most later studies, especially in that his main concern was with the compositional process and how a composer dealt with preexistent models. Furthermore, as exemplified in the first part of his title, "Emulation, Competition, and Homage," he also placed great emphasis on relationships between composers as the inspiration behind the choice of a model.[2] Although his analytical examples came exclusively from chansons, Brown concluded his article by broadly applying his ideas to the composition of masses, and in particular to masses based upon polyphonic cantus firmi.[3] Subsequently, the mass repertoire and specifically the relationships between composers that led to the repeated use of specific cantus firmi and compositional techniques became a favorite focus of borrowing studies.[4]

Without minimizing the influence or important contributions of these studies, it should nonetheless be remarked that in focusing primarily on composer relationships and the technical treatment of borrowed material, it can be too easy to underestimate or overlook the importance of context. This, moreover, is a matter of particular importance for a liturgical genre such as the mass, which served a specific social function. This concern has begun to be addressed, however, by studies that stress the symbolic, extramusical meaning of borrowed material as a primary determinant in cantus firmus selection, a concept akin to basing isorhythmic motets and masses on plainchant cantus firmi because of the liturgical appropriateness of the chant.[5]

In some cases an investigation of a cantus firmus or other preexistent music has led to a greater understanding of a work's context. This can be seen in Michael Long's discussion of Josquin's *Missa di dadi*, in which, through a detailed examination of the secular cantus firmus and other features, Long persuasively argued for a specific ceremonial setting for the mass.[6] Conversely, the specific purpose or occasion for which a mass was written can be invoked to help explain the choice of a model or another unusual borrowing. Rob Wegman, for example, has hypothesized several possible contexts, such as endowments or special occasions, to help explain an unusual mass for St. John that is modeled on the cantus firmus structure of Antoine Busnoys's *Missa L'homme armé*.[7] Though there undoubtedly were situations in which the composer's choice of a model was guided by personal competition or homage, for many masses it seems more fruitful to study borrowing as the inevitable result of the composer's religious and social milieu.

In the present study I shall concentrate on a single mass genre, the polyphonic *Missa de feria*, and unlike the studies cited above I shall examine instances of musical modeling and appropriation that are not necessarily related to cantus firmus material. Within this one genre we can observe surprisingly different aspects of borrowing, and the aspects are all intimately related to the varied contexts for which the masses were written. In one case context shall help illuminate and even determine the presence of borrowing, while in another case musical similarities shall provide valuable insight into the context of the masses.

* * *

The *Missa de feria*, commonly defined as the mass for daily use, may at first seem an unusual choice for this investigation; it may even seem an unusual candidate for polyphonic setting at all. Celebrated on weekdays on which there is no specific feast (that is, on ferial days as opposed to festal days or Sundays), this is the lowest-ranking mass in the Catholic liturgy. As a result, it does not generally contain a Gloria or Credo, and the other chants of the Ordinary sung on these days are among the simplest and shortest plainchant in the repertoire.[8]

Despite this apparent insignificance, however, polyphonic settings of the *Missa de feria* were being written already by the middle of the fifteenth century. As we would reasonably suspect from a mass of low liturgical rank and simple function, most of them are brief, unpretentious, and devoid of contrapuntal complexities; indeed, many of them could very well be examples of improvised polyphony that came to be recorded in mensural notation.[9] Like polyphonic Requiem masses and masses for the Blessed Virgin Mary, these works are properly defined as "plainchant masses" in that instead of having a single cantus firmus throughout the entire mass, each section is a setting of the appropriate item of chant.[10] Moreover, the chant selection seems to have been standardized from the outset: almost every polyphonic setting uses the chants now identified as Sanctus and Agnus XVIII in the *Liber Usualis,* and the most commonly used Kyrie is the one listed as no. 7 (hereafter Melnicki 7) in Margareta Melnicki's catalogue of Kyrie melodies.[11]

The polyphonic ferial masses that have happened to survive (which surely represent only a small fraction of those that were performed during the fifteenth and sixteenth centuries) cover a wide chronological and geographical scope, and as a result, they display a variety of local musical styles and plainchant usage (see Figure 6.1). For example, Gilles Binchois's mass movements and the anonymous English mass are in a three-voice "chanson style" with the chant paraphrased in the superius, a style typical of the mid-fifteenth century.[12] The three anonymous German masses exhibit a "monorhythmic cantus firmus style" that, according to Reinhard Strohm, "remained a hallmark of German plainsong settings (*Choralarbeitung*) for generations."[13] These three masses also share identical plainchant variants, both among themselves and with Ludwig Senfl's later *Missa ferialis;* it is especially noteworthy that they use the same unidentified Kyrie melody.[14] Local chant variants are also evidenced in the ferial masses from Spain and the New World.[15]

In some cases the manuscript sources offer insights into the specific function of these works. For example, several of the masses appear in manuscripts that are devoted almost entirely to polyphony for Holy Week, suggesting that in some locales daily polyphony was reserved for this especially important time. If this was a common observance, then these settings are not necessarily of the lowest liturgical rank after all, as we may have first surmised. The practice of singing polyphonic *Missae de feria* during Holy Week may even have been observed already by the early fifteenth century; not only do our latest examples (especially those of Spanish origin) appear in Holy Week manuscripts, but so also does one of the earliest surviving polyphonic ferial masses, the anonymous English work.

Mass and Sources Dates from *CC*, unless otherwise noted	**Comments**
Kyrie Feriale (#13) & Sanctus/ Agnus (#6)[a] by **Gilles Binchois** (ca. 1400–60) [*TrentC 87; TrentC 92; AostaS D19* **(ca. 1430–40)**]	3-voice "chanson style."
Anonymous untitled mass [*LonBLE 3307* **(ca. 1440)**[b]]	3-voice "chanson style," 3 mvts. Sanctus begins similarly to Binchois's. Manuscript consists primarily of polyphony for Holy Week.
Missa ferialis by **Johannes Martini** (ca. 1430/40–97) [*ModE M.1.13* **(ca. 1481)**; *VatS 35* **(ca. 1487–1503)**; *VerBC 761; 1505[1]*]	4 voices, 3 mvts. Contains material that is imitated in 3 later masses from the papal chapel.
Anonymous untitled mass, probably of French origin [*VatS 63* **(ca. 1495)**[c]]	5 voices, 3 mvts. French indications are its style and the presence of French heraldry in the manuscript.
Three anonymous German masses, titled *Feriale*, *Aliud feriale*, and *Feriale tertii toni* [*WeimB A* **(ca. 1505)**]	4 voices, 3 mvts. Masses share chant variants and exhibit a "monorhythmic cantus firmus style" that was predominant in Germany. Two use an unidentified Kyrie; the third uses Melnicki 7.
Missa de feria by **Matthaeus Pipelare** (ca. 1450–ca. 1515) [*JenaU 21* (Habsburg-Burgundian, **ca. 1521–25**); *1541[1]*]	4 voices, 4 mvts. Contains Gloria and Credo but no Agnus. Canonic and pseudocanonic writing in Gloria and Sanctus may have inspired La Rue's ferial mass. Kyrie seems to be based on both Kyrie XVI and XV.
Missa de feria by **Pierre de la Rue** (d. 1518) [*MechAS s.s.* and *VatS 34* **(ca. 1515–16)**; *JenaU 5* **(ca. 1512–25)**; *MontsM 773* **(by March 1516)**; *MontsM 766* **(ca. 1516–34)**; *BrusBR 15075* **(ca. 1524–34)**]	5 voices, 5 mvts. Each movement is based on a canonic foundation of two voices. Chant models in Kyrie, Sanctus, and Agnus are unidentified.

[a] Identifying numbers from *The Sacred Music of Gilles Binchois.*
[b] Strohm, *The Rise of European Music*, 382–85.
[c] Sherr, *Papal Music Manuscripts*, 9–25 and 220–22.

Fig. 6.1 Polyphonic ferial masses.

Mass and Sources	Comments
Missa de feria by **Antoine de Févin** (ca. 1470–ca. 1512) [*VatS 16* (**ca. 1512–17**); *JenaU 2* (**1512–25**); *VatP 1982* (**ca. 1513–23**); *UppsU 76b* (**ca. 1515**); *JenaU 4* (**ca. 1516–19**); *ToleF 23* (**ca. 1520–35**); *VerBC 760* (**ca. 1520–30**); *CambraiBM 4* (**1526–30**); *'s HerAB 72A* (**1530–31**); *UppsU 76c* (**ca. 1530**); *StuttL 44* (**ca. 1540**); *1516[1]*; *1545[6]*]	5 voices, 5 mvts. Each movement is based on a canonic foundation of two voices, which is most likely modeled on La Rue's *Missa de feria.*
Missa ferialis by **Heinrich Isaac** (ca. 1450/55–1517) [*MunBS 39* (**ca. 1515**); *MunBS 26* (**ca. 1540–47**); *MunBS 33*(**ca. 1550**); *VienNB Mus.18745* (**ca. 1535**)]	4 voices, 3 mvts. Uses unique chants that Edward Lerner has located in a 1488 Gradual from Basel.[d] Several sections (Kyrie I, Pleni, Osanna II) contain 2-voice canons.
Missa ferialis by **Ludwig Senfl** (ca. 1486–1542/43) [*MunBS 37* (**ca. 1510–31**); *MunBS 47* (**ca. 1543–50**)]	5 voices, 3 mvts. Uses same unknown Kyrie as the anonymous German masses. Several sections (Sanctus, Pleni, Benedictus, Agnus I) contain 2-voice canons.
Missa de feria by **Andreas Michot** (fl. 1513–22) [*VatS 55* (**ca. 1515–27**)]	4 voices, 3 mvts. Uses motives from Martini's *Missa ferialis.*
Missa de feria by **Johannes Beauserron** (1475/90–1542) [*VatS 55* (**ca. 1515–27**)]	4 voices, 3 mvts. Uses motives from Martini's *Missa ferialis.*
Missa de feria by **Giovanni Pierluigi da Palestrina** (1525/26–94) [*RISM P664* (**1570**)]	4 voices, 3 mvts. Uses material from both Michot's and Beauserron's *Missae de feria.*
Missa de feria by **Vincenzo Ruffo** (ca. 1508–87) [Ruffo, *Messe a Cinque Voci*, **ca. 1572**; only the 2nd ed. of 1580 (*RISM R3061*) survives]	5 voices, 5 mvts. Uses no apparent chant models.

[d] Isaac, *Opera omnia*, 4: xxx–xxxi.

Fig. 6.1 *(cont.).*

Mass and Sources	Comments
Two anonymous Spanish *Missae feriales* [*MontsM 750*; *GranCG 1* (early seventeenth century)]	4 voices, 3 mvts. Use Spanish chant variants. Both sources contain polyphony for Holy Week.
Five *Missae de feria* from the New World, one (maybe two) by **Pedro Bermúdez** (ca. 1558–ca. 1605), one by **Juan de Padilla** (ca. 1590–1664) [*GuatC 4*;[e] *PueblaC 1*; *ChiN C36* (late sixteenth/early seventeenth century)]	One 5-voice and four 4-voice masses that use the same chant variants as the preceding Spanish masses. All have 3 movements. GuatC 4 consists predominantly of polyphony for Holy Week.

[e] Snow, *A New-World Collection of Polyphony*.

Fig. 6.1 (*cont.*).

Aside from the common chant models, there are very few musical relationships between the various polyphonic ferial masses, nor is there any reason why we should expect to find any in this rather undistinguished genre.[16] By the end of the fifteenth century and well into the sixteenth, however, we can observe the presence of two distinct "families" of ferial masses, one in the preeminent courts of northern Europe and the other at the papal chapel in Rome (see Figure 6.2). Both of these families raise questions about musical borrowing and thus merit close examination.

* * *

At some point in the early part of the sixteenth century, two very unusual *Missae de feria* were composed. One was by Pierre de la Rue, the foremost composer at the Habsburg-Burgundian court from 1492 to 1516, and the other was by Antoine de Févin, an eminent young composer at the French royal court during the reign of Louis XII (r. 1498–1515).[17] A significant feature of these two works is that they both contain all five movements of the Ordinary. This indicates that the masses were appropriate for weekdays of high liturgical rank, for according to the Usage of Paris it was customary to sing the Gloria and Credo during the octaves of the most important feasts of the *Temporale* such as Christmas, Easter, and Pentecost.[18]

In addition to having five movements, these two masses exhibit further striking similarities that set them apart from other fifteenth-century ferial

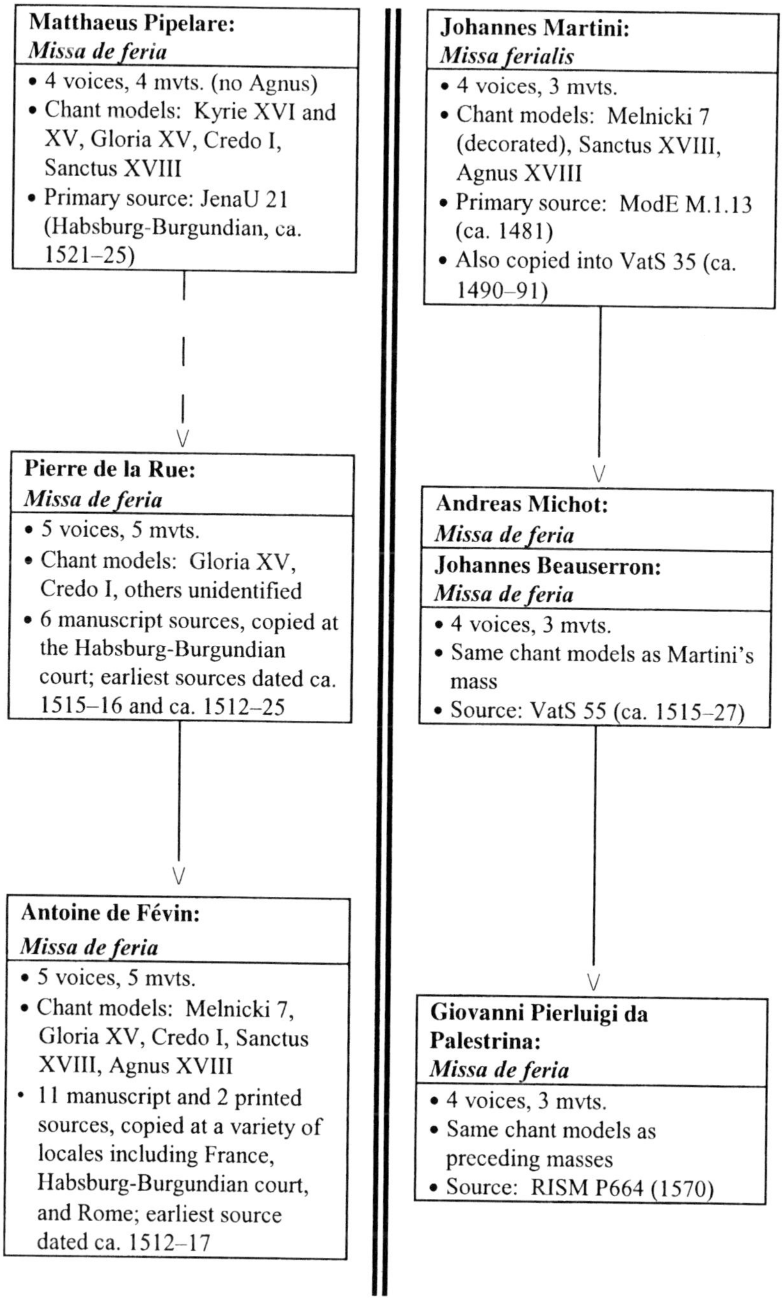

Fig. 6.2 Polyphonic ferial mass families.

Section	***Dux***	***Comes***	**Melodic Interval**	**Rhythmic Interval**
Kyrie I	Tenor 2	Tenor 1	5th	1 semibreve
Christe	Tenor 1	Tenor 2	"	"
Kyrie II	Tenor 2	Tenor 1	"	"
Et in terra pax	Tenor 2	Tenor 1	"	1 breve
Qui tollis	"	"	"	"
Patrem	Tenor 2	Tenor 1	"	2 breves
Et incarnatus est	"	"	"	"
Crucifixus	Tenor 1	Tenor 2	"	"
Sanctus	Tenor 2	Tenor 1	4th	3 breves
Pleni (à2)	—	—	—	—
Osanna	Tenor 1	Tenor 2	"	"
Benedictus (à3)	—	—	—	—
Agnus I	Tenor 1	Tenor 2	5th	4 breves
Agnus II (à2)	—	—	—	—
Agnus III	Tenor 2	Tenor 1	"	"

Fig. 6.3 Canonic structure of La Rue's ferial mass.

masses. For example, both are large-scale, monumental works for five voices, and most unusually, all five movements of both masses are based on a foundation of two voices in strict canon. Despite these similarities, several significant differences make it difficult to assert with confidence that the two masses are directly related. For instance, the treatment of the canon is different in each mass (see Figures 6.3 and 6.4), and whereas Févin based every section of his mass upon the expected plainchants, in all but the Gloria and the Credo of La Rue's mass the chant models remain inexplicably unidentified.[19] Neither mass, moreover, contains any clear, direct melodic quotations from the other. Even in this absence of clear musical parallels, however, it may still be possible to establish that one of these masses is modeled on the other.

We might, for example, follow Howard Brown's lead and "attempt to demonstrate that the relationship between the model and the imitation is significant, that is, that a prominent theme in the model appears in a conspicuous place in the imitation, that the thematic material is individual enough for it to be recognized in a new context, . . . or that some other reason exists to support the claim of a significant relationship."[20] The most likely borrowed element between the two works is unarguably the canonic

Section	*Dux*	*Comes*	Melodic Interval	Rhythmic Interval
Kyrie I	Altus	Superius	4th	3 breves
Christe	"	"	8va	2 breves
Kyrie II	Tenor 1	Tenor 2	5th	3 breves
Et in terra pax	Altus	Superius	4th	3 breves
Qui tollis	Tenor 1	Tenor 2	5th	"
Patrem	Superius	Altus	5th	5 semibreves
Et incarnatus est	Altus	Superius	4th	3 breves
Crucifixus	Tenor 2	Tenor 1	"	"
Et in spiritum	Tenor 1	Tenor 2	5th	5 semibreves
Sanctus	Altus	Superius	4th	3 breves
Pleni (à2)	—	—	—	—
Osanna	Tenor 2	Tenor 1	5th	"
Benedictus (à3)	—	—	—	—
Agnus I	Superius	Altus	"	"
Agnus II (à2)	—	—	—	—
Agnus III (à6)	Tenor 1	Tenor 2	"	"

Fig. 6.4 Canonic structure of Févin's ferial mass.

foundation, and because it occurs in every movement of each mass it does not seem unreasonable to consider it the "prominent theme" of Brown's first criterion. It may also be important that whereas the canon in La Rue's mass appears only in the two tenor voices, in Févin's mass the opening section of every movement presents the canon in the two highest voices. Thus in Févin's mass the canonic voices are more audible and arguably more "conspicuous."

As for Brown's second criterion, that the borrowed material should be "individual," this can be demonstrated clearly enough by just a cursory glance at the other existing *Missae de feria*. Within a genre whose works are marked primarily by simple, possibly even extemporized, counterpoint, the presence of strict canon, which by itself betrays compositional intent, is indeed an unusual and individual characteristic that these two works share. That both masses employ the canon consistently throughout is even more remarkable, and I would suggest that this structural similarity, even if not exact, is in fact an overwhelmingly significant relationship.

Brown's final criterion—"some other reason"—leaves room for further exploration, and here we are aided by Christopher Reynolds, who in another context has stressed the importance of "the chronological and geographical proximity" of the works and their composers.[21] Though the evidence directly connecting our two composers is scant, it is nonetheless significant that they worked at the two most illustrious courts in northern Europe.[22] Indeed, the chapels of lavish secular courts would seem to be the ideal milieux for these grand, monumental ferial masses: not only would the works have played an important role in enhancing the spectacle of religious ceremonies, especially at the most important times of the church year and for other special occasions, but the daily singing of elaborate polyphony by the court chapel could also have served as a status symbol between political rivals.[23]

Ample evidence exists to suggest that the Habsburg-Burgundian and French royal courts did consider La Rue's and Févin's masses in this way. Not only were the two courts vicious political rivals (it must not be forgotten that in 1477 Charles the Bold lost his life and the entire duchy of Burgundy to France in his defeat at the Battle of Nancy), but by the first decade of the sixteenth century each court also had its own reasons for maintaining lavish, ostentatious display for political purposes.[24] The Burgundian court throughout the fifteenth century had been famous for its cultural opulence, but after Charles the Bold's death its political and cultural eminence suffered greatly until Philip the Fair began his reign in 1494.[25] By 1500 Philip had succeeded in restoring the court to its former glory, but his untimely death in 1506 created further problems. Grand public spectacle was undoubtedly of great importance to Philip's sister Marguerite of Austria in her successful attempts to maintain the splendor of the court after 1506. As for the French, John Brobeck has reminded us that grand public display was of great importance to the French court in the early sixteenth century during the investiture of the duchy of Brittany, and also that the years until 1520 were "marked by an unusually large number of state ceremonies involving religious observances."[26]

Even if relations between the courts were far from cordial, the two rivals nonetheless maintained close cultural connections.[27] For example, the respective court poets for each institution, Guillaume Crétin at the French court and Jean Molinet at the Habsburg-Burgundian court, were correspondents, and Molinet even wrote two laments on the death of the famous French court composer Johannes Ockeghem. Alexander Agricola, while working for the Habsburgs, wrote the motet *Transit Anna timor* for the French court in 1505, and works by composers from both courts appear indiscriminately in both Habsburg-Burgundian and French manuscripts.[28] La Rue himself even traveled through France with Philip the Fair on trips to and from Spain in 1501 and 1503, twice singing with the French court chapel (and thus probably with Févin) on the way.

Given this context of two rival courts vying for superior ostentatious display and yet sharing artistic ideas between them, the possibility that

La Rue's and Févin's ferial masses are directly related increases dramatically. While there is still insufficient evidence for postulating a conscious, personal competition between the two musicians themselves, we can gain much by transferring that idea of competition to the desires of their patrons. When viewed from this perspective, there is a much greater probability that one of these composers did in fact use his contemporary's ferial mass as a structural model in a conscious effort to equal or surpass the musical ostentation of the rival court.

The question of priority—that is, who is modeling upon whom—is often a troubling one when considering two works composed within a short time span, and these two *Missae de feria* are no exception. Unfortunately, the manuscript sources offer no assistance, for the earliest ones date from almost precisely the same time. Furthermore, the fact that both masses were apparently very popular and were widely disseminated in a large number of manuscripts significantly diminishes the usefulness of the sources.[29] Nonetheless, an attempt to discover which mass is the older of the two can offer insight into how the competition between patrons affected composers' styles and compositional choices. Indeed, a closer look at the two works yields clues that point to the possibility that Févin borrowed from La Rue.

It is surely significant that of the two composers, La Rue is more noted as a composer of canons. In fact, La Rue is celebrated among the pantheon of Renaissance composers as a master of canonic writing, and among his oeuvre as a whole there are two masses, three motets, and one chanson that are entirely canonic; three masses, two motets, and three chansons that contain a canonic foundation; and five more masses that contain canonic sections.[30] In Févin's known output, on the other hand, only one other work contains a canonic section.[31] Edward Clinkscale has pointed out that the canons in Févin's ferial mass are not always handled in the most adept manner; he comments, for example, that "in the course of any set of entries, the arrival of the canonic voices betokens a deterioration in the ingenuity with which the melodic lines are organized."[32] Clinkscale further remarks that "final cadences are also a problem for him. The canonic voice in this work always frustrates his attempts to create a satisfactory conclusion, and the voice-parts never reach the final note simultaneously."[33] It seems that if one of these two composers were to choose canonic writing as a means of creating an impressive and ostentatious *Missa de feria,* it would be La Rue. Févin, perhaps charged with the task of creating an equally sumptuous work, may have borrowed this structural idea directly from the older composer. His different treatment of the canon could possibly be the result of his not wanting to ape La Rue exactly, that is, of his desire to augment the borrowed concept with his own ideas.

The differences in the treatment of the canon are also significant and support the above hypothesis. La Rue's canon is very carefully structured and sets the foundation upon which his mass is built. The two tenors are consistently the voices that carry the canon, and the melodic interval between them is always a fifth (except in the Sanctus, where it is a fourth). Furthermore, the rhythmic interval between *dux* and *comes* follows a clear pattern throughout the mass: beginning with a semibreve in the Kyrie, it increases to one breve in the Gloria, and each successive movement adds one more breve, concluding with four breves in the Agnus. The canon in Févin's mass, on the contrary, does not exhibit any large-scale, overarching structures; the only noticeable pattern is that every movement begins with the canon in the two highest voices but concludes with it in the two tenor voices. La Rue thus seems much more committed in his mass to integrating the large-scale structure with the canonic foundation, and this adds weight to the possibility that he was the originator of the idea.

There is one more piece of external evidence that supports the hypothesis that La Rue's mass is the older of the two. This comes from the four-voice *Missa de feria* by Matthaeus Pipelare, which also contains a Gloria and Credo.[34] Pipelare is a rather shadowy figure for whom we possess only a sketchy biography. His only documented post, in fact, is as *zangmeester* from 1498 to 1500 for the Illustre Lieve Vrouwe Broederschap, the important Marian confraternity in 's-Hertogenbosch, though Jennifer Bloxam has presented compelling evidence that the composer may also have worked in Ghent.[35] At any rate, his music was definitely known at the Habsburg-Burgundian court, for much of it (including his ferial mass) is preserved in manuscripts copied there.[36]

The most significant feature of Pipelare's mass is that in several sections the composer employs prominent canons and canonlike writing. This happens, for example, throughout the Gloria, where the superius frequently sings in strict imitation with the plainchant in the tenor (though never in true canon). A true canon does occur in the first section of the Sanctus between the tenor and contratenor.[37] It seems quite possible that the canonic foundations in La Rue's and Févin's masses were originally inspired by Pipelare's mass, and La Rue is much more likely to have been familiar with the earlier mass than his younger French contemporary. Not only was La Rue active in the same general locales as Pipelare, but we also have another clear case in which he based a work, his *Missa de Sancto Job,* on a mass by Pipelare, the *Missa de Sancto Livino.*[38]

Although we cannot prove a direct relationship between La Rue's and Pipelare's ferial masses, the confirmed connection between the two composers makes it entirely possible that La Rue expanded Pipelare's use of canon to form the structural foundation of his own monumental ferial mass. Févin could have then borrowed this foundation in a modified form to achieve the same grand effect. Above all, and regardless of the actual order in which the two five-voice masses were composed, what deserves to be stressed is that by concentrating on the historical context and social milieu of the two unusual *Missae de feria* by La Rue and Févin, it has been possible to uncover a wealth of evidence that helps establish the presence of compositional modeling even without the aid of clear melodic parallels.

* * *

In addition to the ferial masses that were composed in northern Europe, there is another distinct and entirely unrelated family of four such masses that originated in Italy, specifically in the Sistine Chapel in Rome. Surprisingly, the earliest of these masses is not by a papal musician but rather by Johannes Martini, who from the early 1470s until his death in 1497 was *maestro di cappella* of Duke Ercole I d'Este's court chapel in Ferrara.[39] Martini composed his *Missa ferialis* no later than 1481, and though he undoubtedly wrote it for the Ferrarese court, it was copied into a Sistine Chapel manuscript by the early 1490s.[40] Martini himself may have introduced the work to the papal chapel during one of his visits to Rome in 1487 and 1488.[41]

The other three masses in this family were written expressly for the papal chapel by musicians singing in the choir. Two of these are by little-known French musicians, Andreas Michot and Johannes Beauserron.[42] The masses appear in a single Cappella Sistina manuscript dated ca. 1515–27, and they were probably not written before 1513, the year the composers first begin appearing in Vatican records.[43] Finally, the latest known ferial mass written for the papal chapel is by Palestrina.[44] This work appears in the composer's third book of masses from 1570, and thus it probably dates from the 1560s, though it also may have originated during Palestrina's brief employment at the Sistine Chapel in 1555.[45]

These masses all contain only three movements, and they are stylistically different from the *Missae de feria* composed in the north. Whereas the northern masses (excluding those by La Rue and Févin) featured primarily homorhythmic textures and simple counterpoint, the Italian masses are considerably more complex contrapuntally and display more rhythmic activity and greater independence among the voices. But what is most remarkable about these four masses, which span a period of at least seventy-five years, is that they all share striking melodic resemblances.[46]

These musical parallels are most obvious in the Kyrie movements. All four masses open with a point of imitation in all of the voices, based on what seems to be a decorated version of Melnicki 7.[47] As is apparent in Figure 6.5, these motives are remarkably alike, and the melodic and rhythmic correspondences are too similar and too lengthy to be mere coincidence.

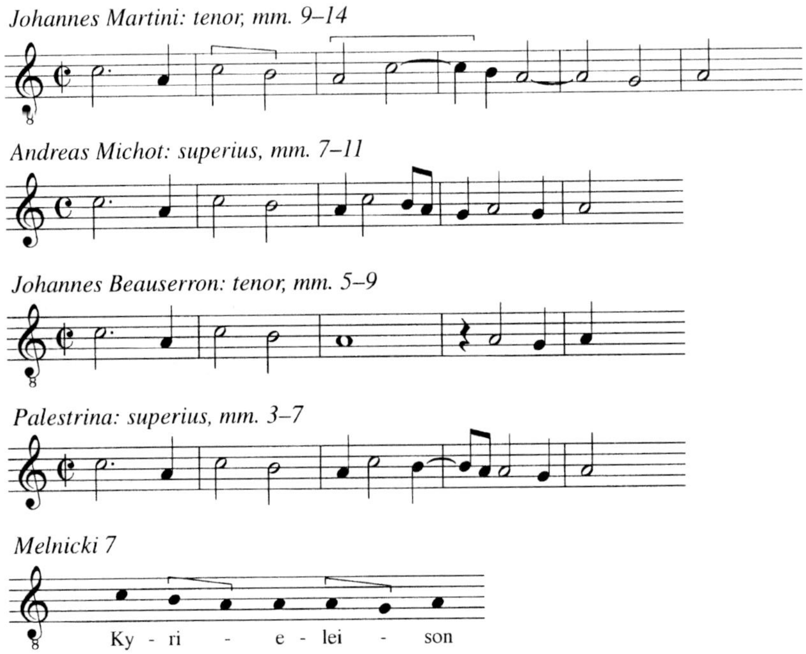

Fig. 6.5 Kyrie motives from four Italian *Missae de feria.*

Beauserron's and Palestrina's masses share another noteworthy relationship in that they both open with another version of this motive, and Palestrina seems to have taken it directly from the earlier mass, even if he did alter the pitches slightly.[48] (The top two voices of both masses are given in Figure 6.6. The bottom two voices are silent in these measures.)

The remaining sections of Palestrina's Kyrie seem to be directly related to the mass by Michot. Both the Christe and second Kyrie of Palestrina's mass are thoroughly imitative sections based on a single motive that migrates constantly among the four voices. Hardly a measure goes by that does not include the motive, and thus the entire section is permeated with this one musical idea. Furthermore, as can be seen in Figures 6.7 and 6.8, Palestrina drew both of these motives from the corresponding sections of Michot's mass. The identical melodic line of the second Kyrie is especially striking because of its length, even though Palestrina altered the rhythm.

Quite unlike the ambiguous relationship between La Rue's and Févin's masses, the melodic similarities within this family are very clear and undoubtedly stem from conscious appropriation on the part of the composers.

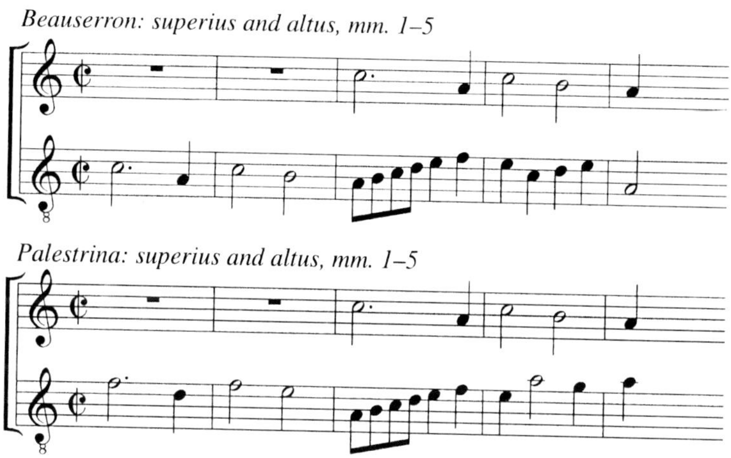

Fig. 6.6 Opening measures of Beauserron's and Palestrina's ferial masses.

But, we might wonder, what is the significance of these relationships? Why, in fact, did the composers borrow? The relative stylistic uniformity among all four masses would seem to effectively rule out competition as a motivating force, and the fact that Martini's mass is unattributed in its Vatican source makes homage unlikely as well.[49] It also seems odd that Palestrina took pains to create ties to works by two relatively unknown composers. What, if anything, can we learn from these examples of borrowing?

The Vatican source that contains Martini's *Missa ferialis*, VatS 35, is the earliest known choirbook compiled by the singers of the choir specifically for their own use. As Adalbert Roth has argued, the contents of this and other early Vatican manuscripts are the result of a conscious attempt at

Fig. 6.7 Christe motives in Michot's and Palestrina's ferial masses.

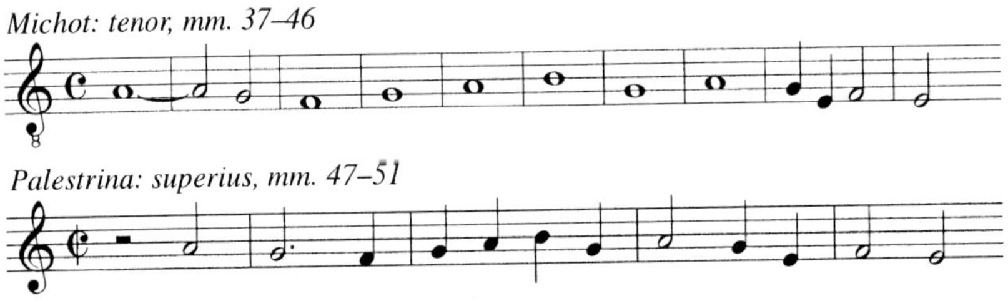

Fig. 6.8 Kyrie II motives in Michot's and Palestrina's ferial masses.

compiling a polyphonic repertoire for the chapel.[50] Jeffrey Dean has further suggested that certain pieces in these manuscripts came to be repeated over a long span of time, thus constituting one of earliest canons in Western music.[51] Because over twenty years after it had been copied Martini's mass was sufficiently familiar to both Michot and Beauserron that they could emulate it, the earlier ferial mass must have been one of the works that had entered into the canon of the Sistine Chapel.

This supposition is strengthened when one recalls that both of the later composers were French musicians who were certainly aware of the styles of the *Missae de feria* sung in the north. In fact, Michot may have even worked alongside Févin at the French court, for the former's name appears in the list of singers taking part in the funeral of Louis XII in 1515.[52] Another Vatican manuscript even contains an anonymous, probably French ferial mass that was copied ca. 1495, so it seems especially significant that Michot and Beauserron apparently overlooked this mass and turned instead to Martini's work when composing their masses for the Italian chapel.[53]

That the two French composers were writing ferial masses at all suggests that this genre reflects another characteristic of the Sistine Chapel's repertoire: the propagation of compositional traditions.[54] It is possible, for example, that other polyphonic *Missae de feria* similar to Martini's were available to Michot and Beauserron and that they also borrowed from these. Whatever the case, it is clear that either the two masses by the French composers or pieces very similar to them had remained staples of the papal repertoire as late as the 1560s, when Palestrina provided yet another similar ferial mass for the chapel's use.

It is most likely that when writing his *Missa de feria* Palestrina did not so much consciously turn to two old masses by obscure, deceased composers but instead drew upon the frequently performed music that was undoubtedly in the ears of the papal choir. It appears, then, that these ferial masses afford us a rare glimpse into a "living repertoire" at the papal chapel throughout the sixteenth century.[55] Whereas in earlier generations composers had based masses and other sacred works on plainchant to give their works authority and to infuse them with meaning, by the sixteenth century these papal musicians were depending upon venerable polyphonic models to achieve the same results.

From the existence of these four masses that clearly borrow from one another, we can now hypothesize with reasonable confidence about an important musical practice at the papal chapel: it was apparently not uncommon for the chants of the Ordinary to be sung daily in polyphony, even if this happened only during the more important times of the liturgical year.[56] This seemingly simple observation is not as self-evident as it may first appear, for despite the great prestige of the papal choir, the role that polyphony played in their services in the late fifteenth and early sixteenth centuries has never been clear.

Although we are fortunate that several constitutions of the chapel from the 1530s and 1540s have survived, as well as diaries and other writings of the papal masters of ceremonies Johannes Burkhard and Paride de Grassis, the information that these sources provide about polyphonic performances is sparse.[57] Writing in 1982, Richard Sherr was able to offer only six firm assumptions about polyphonic practices at the chapel:

1. Singers of polyphony were present at all papal ceremonies.
2. Occasionally, single polyphonic mass movements (particularly Credos) were sung.
3. Motets did not form a regular part of the service, but could be inserted.
4. The Passion was performed with certain sections in polyphony.
5. There were periods of the year when polyphony was banned altogether.
6. The practice of singing a polyphonic or half-polyphonic psalm during Holy Week was new in 1514.[58]

The tendency in recent scholarship has been to downplay the importance of polyphonic music in the regular repertoire of the papal choir and to emphasize instead the continued significance of plainchant. For example, Sherr has proposed that despite the great interest that the papal chapel apparently took in acquiring large-scale mass and motet settings in the early sixteenth century, "the preservation of works in bound manuscripts is not necessarily an indication of their place in a continuous repertory. Indeed, the very act of the compilation of large choirbooks of polyphony may testify to the ephemeral nature of a repertory that was used up very quickly."[59] He further remarks that "the music that the papal choir valued most [such as hymn settings and other simple substitution polyphony for the offices], that is, the music that became a long-standing part of the repertory, is the music we value least. . . . the masses and motets over which we enthuse were precisely the pieces that were considered expendable, constantly to be replaced."[60]

The most extreme version of this stance has been voiced by Jeffrey Dean, who in a controversial article brought to light many negative attitudes toward polyphony that existed within the papal court during the fifteenth and sixteenth centuries.[61] Expanding upon this, he further argued that polyphonic music was primarily the responsibility only of the musicians and that others paid little attention to or possibly did not even care about polyphony.[62] However, the presence in Vatican manuscripts of polyphonic ferial masses that apparently achieved the status of traditional or even canonical works forces us to reconsider this view and remain open to the possibility that daily performances of polyphonic masses were not only frequent but required (and quite possibly appreciated) at the papal court.

Michot's and Beauserron's masses alone might not provoke such considerations, for these two composers were in papal service during the reign of Leo X (r. 1513–21), a pope who was famous in his own day for being an enthusiastic patron of the arts and especially of music.[63] Yet the other two *Missae de feria* by Martini and Palestrina give us reason to assume not only that the singing of daily polyphony was initiated before Leo's reign (thus providing the two French composers with an Italian tradition from which to draw), but also that it continued long enough afterwards to supply Palestrina with venerable polyphonic models for his ferial mass.

* * *

This study has attempted to demonstrate that historical, social, and liturgical context provides the key to understanding two rather different instances of musical borrowing. By factoring context into an examination of the relationships between works of a single genre, we have been able to ascertain with reasonable confidence that two masses with no clear melodic parallels are indeed the products of compositional modeling. Through a study of four other works that do share clear musical parallels it was possible to hypothesize about important aspects of musical practice. Furthermore, in both cases we have seen that even when borrowing practices differ from our traditional views of borrowing in early music, the impulse behind musical modeling and appropriation remained constant for centuries: driving these composers was the desire to enhance the prestige or the meaning of their works by means of references to an esteemed model.

Notes

Earlier versions of this material were presented at the fall meeting of the New England Chapter of the American Musicological Society, September 25, 1999; at Yale University, October 10, 1997; and at the fall meeting of the Southwest Chapter of the American Musicological

Society, October 19, 1996. Preliminary research was made possible through a grant from Rice University, and I am indebted to Honey Meconi for her valuable comments on earlier drafts. For kindly answering letters and providing me with materials, I would also like to thank Jeffrey Dean, Richard Sherr, Murray Steib, and the late Robert J. Snow.

1. Howard Mayer Brown, "Emulation, Competition, and Homage: Imitation and Theories of Imitation in the Renaissance," *JAMS* 35 (1982): 1–48.
2. Brown was influenced by literary studies, in particular G. W. Pigman III, "Versions of Imitation in the Renaissance," *Renaissance Quarterly* 33 (1980): 1–32.
3. This line of thinking had already been initiated by Lewis Lockwood in his "On 'Parody' as Term and Concept in 16th-Century Music," in Jan La Rue et al., eds., *Aspects of Medieval and Renaissance Music: A Birthday Offering to Gustave Reese* (New York: W. W. Norton, 1966; reprint ed., New York: Pendragon, 1978), 560–75; see also Lewis Lockwood, "A View of the Early Sixteenth-Century Parody Mass," in Albert Mell, ed., *Queens College of the City University of New York, Department of Music: Twenty-Fifth Anniversary Festschrift (1937–1962)* (Flushing, N. Y.: Queens College Press, 1964), 53–78.
4. Noteworthy ensuing studies of musical borrowing in masses include Leeman L. Perkins, "The L'Homme Armé Masses of Busnoys and Okeghem: A Comparison," *Journal of Musicology* 3 (1984): 363–96; J. Peter Burkholder, "Johannes Martini and the Imitation Mass of the Late Fifteenth Century," *JAMS* 38 (1985): 470–523; and Murray Steib, "A Composer Looks at his Model: Polyphonic Borrowing in Masses from the Late Fifteenth Century," *TVNM* 46 (1996): 5–41. An important critique of some of the premises guiding these studies is Honey Meconi, "Does *Imitatio* Exist?" *Journal of Musicology* 12 (1994): 152–78.
5. One study of this phenomenon in Guillaume Dufay's masses, though not covering the most famous examples, is Lia Laor, "Concerning the Liturgical Usage of Dufay's Fragmentary Masses," *Current Musicology* 37/38 (1984): 49–58. The process of using chant models to enhance the meaning of a new composition extends even as far back as the twelfth century. Margot Fassler, for instance, has argued that certain Victorine sequences were modeled upon other well-known chants of the same genre to create layers of meaning. See Margot Fassler, *Gothic Song: Victorine Sequences and Augustinian Reform in Twelfth-Century Paris* (Cambridge: Cambridge University Press, 1993), 290–302.
6. Michael Long, "Symbol and Ritual in Josquin's *Missa Di Dadi,*" *JAMS* 42 (1989): 1–22.
7. Rob C. Wegman, "Another 'Imitation' of Busnoys's *Missa L'Homme armé*—and Some Observations on *Imitatio* in Renaissance Music," *JRMA* 114 (1989): 189–202; Wegman, *Born for the Muses: The Life and Masses of Jacob Obrecht* (Oxford: Clarendon, 1994), 213–17.
8. Andrew Hughes, *Medieval Manuscripts for Mass and Office: A Guide to Their Organization and Terminology* (Toronto: University of Toronto Press, 1982), 94–95. The Proper chants that were sung on ferial days were repeated from the immediately preceding feast or Sunday.
9. On the prevalence of improvised polyphony in the later fifteenth century, see especially Rob C. Wegman, "From Maker to Composer: Improvisation and Musical Authorship in the Low Countries, 1450–1500," *JAMS* 49 (1996): 409–79.
10. On plainchant masses, see Edgar H. Sparks, *Cantus Firmus in Mass and Motet 1420–1520* (Berkeley and Los Angeles: University of California Press, 1963; reprint ed., New York: Da Capo, 1975), 94.
11. Margareta Melnicki, *Das einstimmige Kyrie des lateinischen Mittelalters* (Regensburg: Gustav Bosse, 1954), 87. These three chants had been grouped together to form the lowest-ranking service in manuscripts that date as far back as the thirteenth century. For more information on cycles of plainchant masses, see Martin Joseph Burne, "Mass Cycles in Early Graduals: A Study of the Ordinary of the Mass Cycles Found in Medieval and Renaissance Graduals in Libraries in the United States" (Ph.D. dissertation, New York University, 1956); Kurt von Fischer, "Neue Quellen zum einstimmigen Ordinariumszyklus des 14. und 15. Jahrhunderts aus Italien," in Albert van der Linden, ed., *Liber amicorum Charles van den Borren* (Antwerp: Lloyd Anversois, 1964), 60–68; Leo Schrade, "The Cycle of the Ordinarium Missae," in Higini Anglès et al., eds., *In memoriam Jacques Handschin* (Strasbourg: P. H. Heitz, 1962), 87–96; Schrade, "News on the Chant Cycle of the *Ordinarium Missae,*" *JAMS* 7 (1955): 66–69; and Bruno Stäblein, "Messe," *Die Musik in Geschichte und Gegenwart,* 9:155–56 and exx. 2–9. Gustave Reese also explored these chant cycles in "The Polyphonic 'Missa de Beata Virgine' as a Genre: The Background of Josquin's Lady Mass," in *JosCon,* 589–98. For a list of the chant

cycles in the Usage of Paris, which dominated court chapels throughout northern Europe, see Craig Wright, *Music and Ceremony at Notre Dame of Paris 500–1550* (Cambridge: Cambridge University Press, 1989), 82–89.

12. Sparks, *Cantus Firmus,* 51–53. The Binchois movements are edited in Gilles Binchois, *The Sacred Music of Gilles Binchois,* ed. Philip Kaye (Oxford and New York: Oxford University Press, 1992), 62–65 and 88–89. The movements may form a complete mass; see Arthur Parris, "The Sacred Works of Gilles Binchois" (Ph.D. dissertation, Bryn Mawr College, 1965), 1:76–77. The English manuscript is edited in Gwynn S. McPeek and Robert White Linker, eds., *The British Museum Manuscript Egerton 3307* (London: Oxford University Press, 1963). The manuscript is also discussed in Reinhard Strohm, *The Rise of European Music, 1380–1500* (Cambridge: Cambridge University Press, 1993), 383–84.

13. Strohm, *The Rise of European Music,* 526. The masses are not available in modern edition, but musical incipits and a small amount of information are in Karl Erich Roediger, *Die geistlichen Musikhandschriften der Universitäts-Bibliothek Jena,* Claves Jenenses: Veröffentlichen der Universitätsbibliothek Jena 3 (Jena: Walter Biedermann/Frommannsche Buchhandlung, 1935; reprint ed., Hildesheim, Zürich, and New York: Georg Olms, 1985), 202–4.

14. A modern edition of Senfl's mass is in Ludwig Senfl, *Sämtliche Werke,* ed. Walter Gerstenberg et al. (Wolfenbüttel and Zürich: Möseler, 1937–74), 1:55–59. The editors here identify the model as Kyrie XV (ibid., 116), but Roediger claims that it is Kyrie XVIII (*Die geistlichen Musikhandschriften,* 202). Similarities to either chant are slight.

15. Robert J. Snow, ed., *A New-World Collection of Polyphony for Holy Week and the Salve Service: Guatemala City, Cathedral Archive, Music MS 4,* Monuments of Renaissance Music 9 (Chicago and London: University of Chicago Press, 1996), 29–34; see especially 31–33 for a discussion of the ferial plainchants as they were recorded in late-sixteenth- and early-seventeenth-century Roman publications. I am grateful to the late Professor Snow for providing me with photocopies of ferial masses that do not appear in this edition.

16. The most striking relationship is between the Sanctus movements of the Binchois and English masses, the first two phrases of which begin nearly identically in all three voices. It does not seem, however, that one should make too much of these similarities.

17. Modern editions in, respectively, Pierre de la Rue, *Opera omnia,* ed. Nigel St. John Davison, J. Evan Kreider, and T. Herman Keahey, CMM 97 (Neuhausen-Stuttgart: Hänssler/AIM, 1989–), 2:121–74; and Antoine de Févin, *Les Œuvres complètes/Collected Works,* ed. Edward Clinkscale, Gesamtausgabe 11 (Henryville, Ottawa, and Binningen: Institute of Mediaeval Music, 1980–96), 2:80–138. One of the earliest manuscript sources for La Rue's mass is MechAS s.s., dated ca. 1515–16, and one of the earliest sources for Févin's is VatS 16, dated ca. 1512–17. Dates for these and all other manuscripts mentioned are from *CC.*

18. Jennifer Bloxam has cited a thirteenth-century ordinal serving the cathedral of Notre-Dame in Paris (ParisBNL 8884, fol. 293) that specifies the weekdays on which all five movements were sung. See Mary Jennifer Bloxam, "A Survey of Late Medieval Service Books from the Low Countries: Implications for Sacred Polyphony, 1460–1526" (Ph.D. dissertation, Yale University, 1987), 192. Craig Wright has also kindly provided me with a transcription of a thirteenth-century French *Liber cantoris* (SensBM 6, fol. 17) that specifies weekdays on which the Credo is sung. In addition to La Rue's and Févin's masses, a four-voice *Missa de feria* by Matthaeus Pipelare also contains a Gloria and Credo (but no Agnus). A modern edition of Pipelare's mass is in Matthaeus Pipelare, *Opera omnia,* ed. Ronald Cross, CMM 34 (n.p.: AIM, 1966–67), 2:10–22. These masses all use Gloria XV and Credo I.

19. This idiosyncrasy is not so unusual for La Rue. Several other masses by him, such as *Missa Allelulia, Missa de Sancta Anna,* and *Missa de septem doloribus,* have cantus firmi that have not been identified. On two separate occasions J. Evan Kreider has proposed both Agnus XVI and XV as possible chant models for the final movement, but any similarities seem tenuous; see John Evan Kreider, "The Masses for Five and Six Voices by Pierre de la Rue" (Ph.D. dissertation, Indiana University, 1974), 1:41, and La Rue, *Opera omnia,* 2: lxii. The chant models in Févin's mass, which are at times so highly paraphrased as to be practically unrecognizable, are Melnicki 7, Gloria XV, Credo I, Sanctus XVIII, and Agnus XVIII. This identification is disputed by Edward Clinkscale, who has proposed Kyrie XVIII, Sanctus XVII, and Agnus XV as possible models; see Févin, *Collected Works,* 2:xvi.

20. Brown, "Emulation, Competition, and Homage," 14.

21. Christopher A. Reynolds, "The Counterpoint of Allusion in Fifteenth-Century Masses," *JAMS* 45 (1992): 228–60, at 247. Reynolds's main concern in this article is to establish the presence of allusions to secular works in selected masses.
22. Both courts, furthermore, followed the Usage of Paris. A Habsburg-Burgundian court ordinance dated January 1, 1469 (n.s.), that was subsequently reissued in 1497 and 1500 states that "High Mass will be celebrated daily in the chapel 'in chant and discant,' according to the use of Paris, as is the custom." This translation is from Martin Picker, "The Habsburg Courts in the Netherlands and Austria, 1477–1530," in Iain Fenlon, ed., *The Renaissance: From the 1470s to the End of the 16th Century* (London: Macmillan, 1989), 216–42, at 218. The original document is transcribed in David Fallows, "Specific Information on the Ensembles for Composed Polyphony, 1400–1474," in Stanley Boorman, ed., *Studies in the Performance of Late Medieval Music* (Cambridge: Cambridge University Press, 1983), 109–59, at 145–59.
23. This was first suggested in Bloxam, "A Survey of Late Medieval Service Books," 191–95.
24. On Charles the Bold's defeat in 1477, see Richard Vaughan, *Charles the Bold: The Last Valois Duke of Burgundy* (London: Longman, 1973), 399–432, and Vaughan, *Valois Burgundy* (London: Penguin, 1975), 194–227.
25. For more details about musical patronage at the Habsburg-Burgundian court, see Vaughan, *Valois Burgundy,* 162–93, and Picker, "The Habsburg Courts." The classic article about the court chapel is G[eorges] Van Doorslaer, "La Chapelle musicale de Philippe le Beau," *Revue belge d'archéologie et d'histoire de l'art* 4 (1934): 21–57, 139–65.
26. John T. Brobeck, "Musical Patronage in the Royal Chapel of France under Francis I (r. 1515–1547)," *JAMS* 48 (1995): 187–239, at 212–13. Further information about the French court during earlier periods can be found in Leeman L. Perkins, "Musical Patronage at the Royal Court of France under Charles VII and Louis XI (1422–83)," *JAMS* 37 (1984): 507–66; Frank Dobbins, "Valois," *New Grove II;* and Richard Sherr, "The Membership of the Chapels of Louis XII and Anne de Bretagne in the Years Preceding Their Deaths," *Journal of Musicology* 6 (1988): 60–82.
27. The evidence in this paragraph is from Honey Meconi, "French Print Chansons and Pierre de la Rue: A Case Study in Authenticity," in Jessie Ann Owens and Anthony M. Cummings, eds., *Music in Renaissance Cities and Courts: Studies in Honor of Lewis Lockwood* (Warren, Mich.: Harmonie Park Press, 1997), 187–214, at 192–94.
28. Unfortunately, the dearth of surviving French royal manuscripts from this period precludes knowing the full extent to which music by Habsburg-Burgundian composers was known and performed at that court.
29. La Rue's mass appears in six manuscripts, all of which were copied at the Habsburg-Burgundian court (though several were sent to other locales such as the papal court and Spain). Févin's work appears in thirteen sources, which were copied in a variety of locales including Cambrai, Stuttgart, and the papal chapel. All sources are listed in Figure 6.1.
30. Honey Meconi, "La Rue, Pierre de," *New Grove II,* 284–85. If one includes anonymous chansons attributed to La Rue by modern scholars, this adds one additional entirely canonic work and five more chansons composed over canonic foundations. I am grateful to Professor Meconi for sharing a copy of her article with me prior to its publication.
31. This is his *Missa super O quam glorifica luce,* which contains a mensuration canon in the Benedictus.
32. Edward Henry Clinkscale, "The Complete Works of Antoine de Févin" (Ph.D. dissertation, New York University, 1965), 1:69.
33. Ibid. The most glaring examples occur in the final measures of Kyrie I, Et incarnatus est, and Agnus III. In all of these cases (plus others), several voices reach their final notes individually.
34. This mass is cited in note 18 above. Ronald Cross has pointed out similarities between Pipelare's *Missa de feria* and the five-voice *Missa sine nomine* by Ockeghem that contains only a Kyrie, Gloria, and Credo. Pipelare's chant models are Kyrie XVI and XV, Gloria XV, Credo I, and Sanctus XVIII, and because Ockeghem's movements are based on Kyrie XVI, Gloria XV, and Credo I, Cross labels this earlier work a ferial mass. Citing compositional similarities between the two works, Cross even proposes the existence of a common practice in ferial mass composition, though with scant supporting evidence from other works; see Ronald Cross, "Matthaeus Pipelare: A Historical and Stylistic Study of His Works" (Ph.D. dissertation, New York University, 1961), 1:150–57. Modern editions of Ockeghem's mass are available in Johannes Ockeghem, *Collected Works,* 3 vols., 2nd ed. of vols. 1–2, ed. Dragan Plamenac and Richard Wexler, American Musicological Society Studies and Documents 1,

3, 7 (n.p.: American Musicological Society, 1959–92), 2:77–79, and also Ockeghem, *Masses and Mass Sections,* ed. Jaap van Benthem (Utrecht: Koninklijke Vereniging voor Nederlandse Muziekgeschiedenis, 1994–), I/3: 11–23.

35. M. Jennifer Bloxam, "In Praise of Spurious Saints: The *Missae Floruit egregiis* by Pipelare and La Rue," *JAMS* 44 (1991): 163–220.

36. For a list of Pipelare's works and their sources, see Ronald Cross, "The Life and Works of Matthaeus Pipelare," *Musica Disciplina* 17 (1963): 97–114.

37. There are other *Missae de feria* contemporary to or slightly later than La Rue's and Févin's that make use of some strict imitation and occasional rudimentary canons, including the ferial masses by Isaac, Senfl, and even the first of the three anonymous German masses. Clearly the very simple ferial plainchant melodies invited imitative manipulations, and we cannot and should not attempt to discern a thread of influences in all of these works. No other masses, however, employ canon to such an extent and in such a monumental fashion as those by La Rue and Févin, and the proximity of La Rue and Pipelare is undoubtedly worth noting. An edition of Isaac's *Missa ferialis* is in Heinrich Isaac, *Opera omnia,* ed. Edward R. Lerner, CMM 65 (n.p.: AIM, 1974–), 4:114–26.

38. Bloxam, "In Praise of Spurious Saints." Bloxam points out more probable borrowings and connections between the composers on pp. 210–11.

39. Information about Martini and his importance to the musical life at the Ferrarese court is in Lewis Lockwood, *Music in Renaissance Ferrara 1400–1505: The Creation of a Musical Center in the Fifteenth Century,* Studies in the History of Music 2 (Oxford: Clarendon; New York: Oxford University Press; Cambridge, Mass.: Harvard University Press, 1984), 167–72.

40. The main source of the work is the Ferrarese manuscript ModE M.1.13, which is dated ca. 1481. The Vatican manuscript in which it appears is VatS 35, dated ca. 1487–90. A modern edition of the mass is available in Johannes Martini, *Masses,* ed. Elaine Moohan and Murray Steib, Recent Researches in the Music of the Middle Ages and Early Renaissance 34–35 (Madison: A-R, 1999), 34:57–76. I am grateful to Professor Steib for kindly providing a transcription of the mass prior to its publication.

41. Lockwood, *Music in Renaissance Ferrara,* 170.

42. Biographical information for these composers can be found in Nors S. Josephson, ed., *Early Sixteenth-Century Sacred Music from the Papal Chapel,* CMM 95 (Neuhausen-Stuttgart: Hänssler/AIM, 1982), ix–x. For further information on Michot, see Josephson, "Michot, Andreas," *New Grove II,* and Brobeck, "Musical Patronage," 197 and 213–14. On Beauserron, see Joshua Rifkin, "Beausseron, Johannes Bonnevin alias," *New Grove II.*

43. The manuscript is VatS 55. Modern editions of both works are available in Josephson, *Early Sixteenth-Century Sacred Music,* 82–93 and 347–58.

44. Modern edition available in Giovanni Pierluigi da Palestrina, *Le opere complete,* ed. Raffaele Casimiri et al. (Rome: Istituto Italiano per la storia della musica, Edizione Scalera, 1939–99), 6:84–96.

45. Lewis Lockwood, Noel O'Regan, and Jessie Ann Owens, "Palestrina, Giovanni Pierluigi da," *New Grove II,* 939. Palestrina continued to write works for the papal chapel as late as 1565.

46. There is one other sixteenth-century Italian ferial mass that deserves to be mentioned, even though it is unrelated to the masses composed for the papal chapel: the five-voice *Missa de feria* by Vincenzo Ruffo. This work appears in the composer's published edition of five-voice masses, which probably dates from 1572 (only the second edition of 1580 survives), and it was most likely written during Ruffo's employment at the Cathedral of Milan from 1563 to 1572; see Lewis Lockwood, *The Counter-Reformation and the Masses of Vincenzo Ruffo* (Venice: Fondazione Giorgio Cini, [1970]), 239–40. The mass contains all five movements of the Ordinary, and unlike any other surviving *Missa de feria,* it is entirely freely composed and contains no apparent references to chant. In fact, with its diminutive proportions and lack of models it closely resembles the sixteenth-century genre of the *Missa brevis,* an intriguing connection that may shed some light on that rather mysterious genre. Ruffo's *Missa de feria* is not available in modern edition in its entirety, but the Credo is edited in Luigi Torchi, ed., *L'Arte musicale in Italia* (Milan: Ricordi, n.d.), 1:197–204.

47. It is difficult to confirm that the chant model is indeed Melnicki 7 because the two graduals in the Cappella Sistina collection, VatS 5 and VatS 12, do not include the chants for the ferial mass in their kyriales; the lowest-ranking mass in both is "In festis minoribus simplicibus."

However, the Vatican library does possess a notated missal of seemingly fourteenth-century Italian origin (VatRE 2049) whose kyriale contains a mass "In ferialibus diebus" that consists of Melnicki 7 (undecorated), Sanctus XVIII, and Agnus XVIII (fol. 16). Clues about the origin of this missal come from the hand and from the fact that the calendar in the opening folios contains many feast days devoted to relatively unknown popes. A printed Roman gradual from 1499 also includes Melnicki 7 together with Sanctus XVIII and Agnus XVIII under the rubric "In ferialibus diebus;" see *Graduale secundum morem sancte Romane ecclesie* (Venice: Antonius de Giunta, 1499), fol. 346v–347r. The ferial Kyrie in this print adds one interior pitch to the "standard" melody of Melnicki 7, but it still does not resemble that in the polyphonic Italian *Missae de feria.* I am grateful to David Rothenberg for checking this source for me.

48. Nors Josephson has labeled Palestrina's Kyrie I a parody of Beauserron's corresponding section (*Early Sixteenth-Century Sacred Music,* x). Although Josephson also notes the similarities between Beauserron's and Michot's masses in "Michot," 601, he nowhere mentions the further striking resemblances between Palestrina's and Michot's masses. Incidentally, the *Missa de feria* does not fit with other Palestrina masses usually considered parody masses; see Quentin W. Quereau, "Aspects of Palestrina's Parody Procedure," *Journal of Musicology* 1 (1982): 198–216.
49. As an important caveat, it should be mentioned that Martini's *Missa ferialis* began circulating with an attribution to Josquin in Petrucci's *Fragmenta missarum* (1505[1]). Thus later papal musicians may have believed they were borrowing from that renowned musician.
50. Adalbert Roth, "Liturgical (and Paraliturgical) Music in the Papal Chapel towards the End of the Fifteenth Century: A Repertory in Embryo," in Richard Sherr, ed., *Papal Music and Musicians in Late Medieval and Renaissance Rome* (Oxford: Clarendon, in association with the Library of Congress, Washington, D.C., 1998), 125–37, esp. 131–32.
51. Jeffrey Dean, "The Evolution of a Canon at the Papal Chapel: The Importance of Old Music in the Fifteenth and Sixteenth Centuries," in Richard Sherr, ed., *Papal Music and Musicians in Late Medieval and Renaissance Rome* (Oxford: Clarendon, in association with the Library of Congress, Washington, D.C., 1998), 138–66.
52. For this list, on which Michot appears as "Michau," see Brobeck, "Musical Patronage," 197. The only wrinkle to this idea is that Michot's name begins appearing in papal documents in 1513 (Josephson, *Early Sixteenth-Century Sacred Music,* ix). Josephson, "Michot," makes no mention of employment in France. Beauserron was a member of the Sainte-Chapelle in Paris from 1511 to 1514 (Rifkin, "Beausseron," 30).
53. The manuscript, VatS 63, is discussed in detail in Richard Sherr, *Papal Music Manuscripts in the Late Fifteenth and Early Sixteenth Centuries,* Renaissance Manuscript Studies 5 (Neuhausen-Stuttgart: Hänssler/AIM, 1996), 9–25, 201–10, and 220–22. Although it cannot be stated with certainty that the mass is a northern work, evidence suggests that it is by a French composer. For example, the manuscript contains French heraldry in its earliest layer (of which the ferial mass is a part), which most likely connects it to Charles VIII's visit to Rome in 1495. Other pieces included in the early layers of copying are by northern musicians, including two masses attributed in the manuscript to Ockeghem. The ferial mass itself, furthermore, is stylistically more similar to other northern masses than to the works composed in Italy.
54. Mitchell P. Brauner, "Traditions in the Repertory of the Papal Choir in the Fifteenth and Sixteenth Centuries," in Richard Sherr, ed., *Papal Music and Musicians in Late Medieval and Renaissance Rome* (Oxford: Clarendon, in association with the Library of Congress, Washington, D.C., 1998), 167–74.
55. I borrow this phrase from Jessie Ann Owens, who offered it in response to this paper at the fall meeting of the New England Chapter of the American Musicological Society, Brandeis University, September 25, 1999.
56. Richard Sherr has noted that almost every Vatican manuscript that contains *Missae de feria* also contains the motet *Domine, non secundem* (the tract for Ash Wednesday) in close proximity to the daily mass. He concludes that this was done for liturgical reasons, for "a *Missa de feria* was the only type of mass that could be performed on Ash Wednesday" because of its omission of the Gloria and Credo. He also cites a diary notation in which papal master of ceremonies Johannes Burkhard complained that on Ash Wednesday 1495 the Tract and the Sanctus had both been sung "in figuris." See Richard Sherr, "*Illibata Dei Virgo Nutrix*

and Josquin's Roman Style," *JAMS* 41 (1988): 434–64, at 461. This is an astute observation, yet whereas it clarifies the liturgical role of the motet and provides an important clue about when polyphonic *Missae de feria* may have been performed at the papal chapel, we should not conclude that polyphonic daily masses were sung only on Ash Wednesday. One is reminded of the northern manuscripts that seem to indicate that polyphonic ferial masses were performed during Holy Week.

57. On these sources and available modern editions, see especially Jeffrey Dean, "Listening to Sacred Polyphony *c.* 1500," *Early Music* 25 (1997): 611–36; Richard Sherr, "The Singers of the Papal Chapel and Liturgical Ceremonies in the Early Sixteenth Century: Some Documentary Evidence," in P. A. Ramsey, ed., *Rome in the Renaissance: The City and the Myth, Papers of the Thirteenth Annual Conference of the Center for Medieval and Early Renaissance Studies* Medieval & Renaissance Texts and Studies 18 (Binghamton, N.Y.: Center for Medieval and Early Renaissance Studies, 1982), 249–64; and Sherr, "The Papal Chapel ca. 1492–1513 and Its Polyphonic Sources" (Ph.D. dissertation, Princeton University, 1975), 2–120.
58. Sherr, "The Singers of the Papal Chapel," 258. More details about polyphonic passions at the papal chapel are in Sherr, "The 'Spanish Nation' in the Papal Chapel, 1492–1521," *Early Music* 20 (1992): 601–9, at 602–4. Our knowledge of the role that motets played in papal services has been augmented by Anthony M. Cummings, "Toward an Interpretation of the Sixteenth-Century Motet," *JAMS* 34 (1981): 43–59.
59. Richard Sherr, "Preface," in Richard Sherr, ed., *Papal Music and Musicians in Late Medieval and Renaissance Rome* (Oxford: Clarendon, in association with the Library of Congress, Washington, D.C., 1998), ix–xiii, at xi.
60. Ibid.
61. Dean, "Listening to Sacred Polyphony."
62. Dean softened and clarified several of his points in an ensuing communication with Honey Meconi; see "Listening to Sacred Polyphony," *Early Music* 26 (1998): 375–79 and 537–39.
63. The number of musicians in the papal chapel from the 1480s until Leo X's reign had ranged between twenty and twenty-five, and during Leo's pontificate the chapel increased to thirty-six members in addition to an extensive private chapel. For further information on musical patronage under this first Medici pope, see André Pirro, "Leo X and Music," trans. Gustave Reese, *Musical Quarterly* 21 (1935): 1–16; Herman-Walther Frey, "Regesten zur päpstlichen Kapelle unter Leo X. und seiner Privatkapelle," *Musikforschung* 8 (1955): 58–73, 178–99, 412–37; 9 (1956): 46–57, 139–56, 411–19; Robert Steiner, "Von Leo X bis Clemens VII: Zwei Humanisten und ein Puritaner auf dem päpstlichen Thron und ihr Verhältnis zur Kunst," in Victor Ravizza, ed., *Festschrift Arnold Geering* (Bern and Stuttgart: Paul Haupt, 1972), 185–99; Sherr, "The Papal Chapel," 78–80; and Bonnie J. Blackburn, "Music and Festivities at the Court of Leo X: A Venetian View," *Early Music History* 11 (1992): 1–37. During his reign, Leo X even acquired the ferial masses by both La Rue and Févin, which appear respectively in VatS 34 and VatS 16.

7

Mid-Sixteenth-Century Chanson Masses: A Kaleidoscopic Process

CATHY ANN ELIAS

Scholars have created theories to describe the origin, process, and evolution of the imitation mass (also known as the parody mass) from the fourteenth to sixteenth centuries.[1] Often discussions focus on two points: (1) how much material is borrowed—one, two, or all of the voices of the model—and (2) the style of the model. For instance, Ludwig Finscher describes the development of the imitation mass as an evolutionary process spanning two centuries. His theories concentrate on the progressive, additive quality of the amount of borrowing from the model, focusing on masses from the fifteenth and sixteenth centuries. To illustrate this process, he creates discrete categories based on specific types of borrowing evolving over time.[2] Lewis Lockwood suggests that a change in the type of model used for these masses around 1500 was the catalyst that led to the full-fledged imitation mass, rather than there being an evolutionary process based on the amount borrowed from the polyphonic model. This new style model, the four-part imitative motet, grew out of a process of simultaneous composition: individual lines of text were set in somewhat discrete imitative complexes or in short homorhythmic passages in a simultaneous manner, rather than in the old-fashioned line-by-line linear style. He concludes that a composer using such a model was forced to borrow individual motives and vertical units rather than single lines.[3] These theories became the main criteria used to determine and understand the development leading to the mature imitation mass.

Although scholars agree that the imitation mass reached its maturity by the middle of the sixteenth century, we still lack a workable definition that accounts for the variety of techniques and borrowing practices employed simultaneously within masses from this period. A close reading of mid-sixteenth-century chanson masses will show that theories of an evolutionary development of the genre and categorization according to type, such as masses based on a cantus firmus or points of imitation, are in need of revision.[4] The imitation mass should not be viewed as the result of a long evolutionary process based on the amount or type of material borrowed from a model or preexistent tune, not only because borrowing has always played an important role in music but also because, as I will show, older procedures are used side by side with newer ones in these masses and the amount of borrowed material varies greatly from one imitation mass composer to another.

The questions I would like to address in this essay are not how much or how little a composer borrowed, but rather what techniques he used, how he used them, and what effect the style of the model had on the mass. Most importantly, how did the composer incorporate new and borrowed material to create a work in his own style? In what fashion did he transform and integrate the model material with his newly composed music to develop a personal style that both differs from and embraces aspects of the model's style?

As a basis for this study, I will discuss mid-sixteenth-century chanson masses because they display more interesting compositional designs and combinations of techniques than masses based on motets from the same period, and because they have been studied the least. I will show that many of the techniques and types of borrowing associated with early or late imitation masses and the discrete stages of development that supposedly occur within the imitation mass over the course of a century or more are all present in mid-sixteenth-century chanson masses. For instance, the significant difference between Gombert and Palestrina is not how much or in what manner they borrowed material from a model in composing masses—as one might expect from theories about borrowing—but rather the way in which each man developed a personal style based on his reuse of the chanson material and the individualistic compositional decisions and considerations he made.

Pervading imitation and seamless polyphony, most prevalent in sacred music, was the pan-European style by the second quarter of the sixteenth century. This style is characterized by the simultaneous use of imitative and nonimitative lines set in a continuous overlapping fashion. During this period, a new organizing procedure for writing masses became popular with many composers, including Nicolas Gombert, Clemens non Papa, Adrian Willaert, Thomas Crecquillon, Jean Richafort, Claudin de Sermisy, and Pierre de Manchicourt. The individual techniques used in composing music in this new fashion—such as cantus firmus, ostinato, and paraphrase methods—were not novel, but were woven together in unique ways, eventually transforming and superseding older conventions. The shift that eventually occurred within this general polyphonic fabric is from the predominance of cantus firmus and paraphrase techniques used in a cyclic fashion to a new, more integrated and varied compositional process that reaches its maturity in the composers named above. Among them, only Clemens, Gombert, Crecquillon, and Manchicourt wrote masses based on chansons.

For Clemens, Gombert, Manchicourt, and Crecquillon, this compositional shift resulted in a new, more musically integrated approach to setting the mass text. In their chanson masses based on a multivoiced preexistent piece (or pieces), they used the model in a variety of different ways with no particular method predominating, and with no single device or technique controlling the entire structure. This new procedure is simply the combination of all the previously used contrapuntal devices, now reinterpreted and presented in a fragmented, concentrated fashion, transforming the imitative texture and structure within a familiar context, the polyphonic setting of the mass text. Consequently, cantus firmus technique, previously employed as a precompositional, schematic, structural device for creating and organizing a cyclic mass, now simply is one of the many different ways to structure material within parts of the mass, having no greater long-term structural importance than any of the other methods employed. Paraphrase too, used as a large-scale unifying device, now becomes just one of the many short-term procedures in an imitation mass. This compositional shift—using a variety of techniques on a local level—is a way of writing where a sense of long-term development, hierarchy, and balance gives way to a style that often eludes clear-cut spatial articulation. These masses are no longer based on a predetermined long-range structural plan: the asymmetric juxtaposition of elements replaces hierarchical relationships, and the borrowed material becomes an audible but undifferentiated part of the fabric's texture.

The integration of the newly composed counterpoint with borrowed material was achieved by constant rhythmic modification of the latter, causing the polyphony to conform to specific harmonic considerations: by changing the rhythm one can realign the vertical sonorities. Although it is clear that the composers were setting lines of a text, often too little attention was given to text declamation. The new texture was articulated and clarified by the reiteration of an unpredictable set of unfolding variants of borrowed material and their modified vertical counterpoint of different lengths and dimensions derived from the model. The outcome is a work with a large-scale sense of homogeneity and a cohesive style. The audible musical coherence is the result of continuous reiteration of variants of a familiar chanson. This unceasing shift of borrowed material throughout these chanson masses is much like the homogeneous but ever-changing, unpredictable patterns produced by a kaleidoscope: this process inherently makes long-term, predetermined, hierarchical structures impossible.

In describing a compositional process, it is not historically justifiable to isolate and list individual techniques because compositional process rarely involves stringing together a linear progression of isolated phenomena to make a composite work. For my analysis, however, such a list will be helpful in order to summarize the diverse elements that make up this new style and to explain the type of cohesion that resulted. The order I have selected to list these techniques should be viewed as arbitrary, chosen for clarity and convenience: no particular device is necessarily used as a precursor to any other one, and the presentation is not meant to describe an evolutionary process.

The understanding of any genre involves defining it by means of common elements shared by all examples. The result is an incomplete and inexact scaffold that only could reflect the worst piece in the genre, the one that lacks all originality and imagination. But unfortunately it is exactly this outline that we need to begin with so that we can say why each work is different from it, and it is this difference that will—in the end—define an individual composer's style. For example, in studying works in sonata form by Haydn, Mozart, and Beethoven, we focus on why their pieces do not fit the mold, and study the innovative ways they reuse a familiar outline. We will follow the same procedure for mid-sixteenth-century chanson masses. A variety of techniques found in the chanson masses of Clemens, Gombert, Manchicourt, and Crecquillon appear and reappear, validating them as entities worth codifying and discussing. By themselves, they are nothing more than the most common elements at a composer's fingertips, but once contextualized and repeated within a particular composer's works, they will provide the clues to the unique aspects of that individual's style and compositional process. In fact, what will become more important is not what they borrowed but how they integrated the borrowings with their own material to create a unique, cohesive style. Consequently, the techniques I outline below can be considered the pan-European techniques associated with these and other composers' imitation masses constructed from chansons. Within these composers' chanson masses, the following techniques are used with borrowed material and combined freely with new counterpoint in a continuous, fragmented, varied fashion:

1. Cantus firmus: (a) using an entire voice from the model throughout a section or a *single* movement of the mass or (b) borrowing an

entire voice from the model and either rearranging the phrases or interpolating new counterpoint between them, then using it within a movement

2. Ostinato and derived techniques: (a) using an ostinato figure in a schematic fashion throughout a small section or (b) reiterating a motive throughout a passage in a free style, saturating the texture with its presence
3. Motivic rescaffolding: borrowing material to construct a new scaffold that has no structural relationship to the model
4. Partial scaffolding: borrowing several but not all of the voices from the model in their exact or relative structural position (switching two voices but keeping the vertical relationship between them intact), and adding new counterpoint to this frame
5. Block structuring: borrowing all of the voices from the model in their exact or relative position
6. Block restructuring: borrowing all of the voices from the model and reworking the complex
7. Block manipulation: borrowing an entire complex and joining it to another one from the model; thus, nonconsecutive blocks of material are taken from the model and juxtaposed in the mass
8. Block interpolation: borrowing a passage from the model and interpolating newly composed blocks periodically throughout it
9. Varied block reiteration: borrowing a block of material, repeating it, and slightly modifying it to accommodate a desired compositional plan

In this essay I will discuss and illustrate four of these techniques: cantus firmus, block interpolation, partial scaffolding, and varied block reiteration. Unlike the other well-known techniques on this list, these are less commonly associated with the mid-sixteenth-century chanson, motet, or madrigal mass. When presented in a larger musical context, these selected techniques reveal new insights into this integrated compositional procedure.

Both Manchicourt and Gombert fragmented and internalized their cantus firmi: instead of using the borrowed melody as an external, precompositional, large-scale structural device (as in the cyclic masses of the previous generations), they integrated it into the polyphonic fabric only in selected sections and movements of their chanson masses. For instance, Gombert used the entire top voice of the model as the superius in an Agnus section in each of his three chanson masses (in *Missa Fors seulement,* the music for the Kyrie and Agnus is the same, so he used cantus firmus technique for two movements instead of only one, as in the other two masses). Consequently, the cantus firmus is not used as the the dominant technique for organizing all movements in this new type of cyclic mass. Instead, cyclic unity is achieved by repeating varied fragments from the model. A borrowed fragment may consist of an entire melody, a complex of voices, a single phrase, or a motive. This results in an audible musical coherence based on a familiar but unpredictable, varying structure, unlike that of the cantus firmus mass, where the recurring borrowed melody often is presented throughout the mass in at least a predictable if not audible schematic fashion.

The chansons Gombert selected for these three masses illustrate his indifference to a model in a particular style. *Missa Je suis desheritée* is based on a Parisian-type chanson by Cadéac, *Missa Sur tous regrets* is based on a Franco-Flemish chanson by Richafort, and *Missa Fors seulement* is based on two models composed in a linear fashion by composers of the previous generation, Pipelare and Févin.[5] Despite the diversity of model styles, Gombert had no problem using complexes of borrowed material throughout these masses. He also appears to have had no difficulty using the entire top voice of these models as the superius in an Agnus section in each of these masses, and the Kyrie movement too in the case of *Missa Fors seulement.*

For instance, the superius of Cadéac's *Je suis desheritée* is characterized by clear-cut phrases, square rhythms, syllabic text setting, and harmonic clarity, whereas the superius of Févin's *Fors seulement* is characterized by long, florid, irregular phrases. Both of these melodies have little in common with each other or with Gombert's style. When these melodies are surrounded with the web of carefully set imitative and nonimitative overlapping lines of varying lengths so characteristic of Gombert's music, what stylistic identity they initially retained—removed from their original context— is now obscured, and they become an integrated part of the contrapuntal fabric (see Figures 7.1a, 7.1b, 7.2a, and 7.2b).[6]

The constructive principle followed in these two Agnus settings is similar: most of the counterpoint added to the borrowed top lines is newly composed, although in each case Gombert also created a few new points of imitation derived from chanson motives, occasionally in their original structural position. The first bracketed phrase in the two bass voices is a motive based on a variant of the theme, since it rises to a fifth rather than to a fourth. Because Gombert typically composed phrases of varying lengths that continuously overlap, he only had to adjust his style slightly to accommodate the short, discrete phrases of Cadéac's top voice used in the Agnus of *Missa Je suis desheritée* (see Figure 7.1b). For instance, in the first seven measures, he set Cadéac's short opening superius phrase in imitation in the lower voices, while using the same borrowed phrase as the beginning of much longer lines, more characteristic of his own style, in two of the upper parts. In each of these instances the chanson phrase is set syllabically, thus articulating it and adding musical unity to the passage. In the remaining five

Fig. 7.1a Cadéac, *Je suis desheritée,* mm. 1–13.

Fig. 7.1b Gombert, *Missa Je suis desheritée,* Agnus Dei, mm. 30–42 (cantus firmus technique); transcription after Nicolas Gombert, *Opera omnia,* 11 vols., ed. Joseph Schmidt-Görg, CMM 6 (Rome: AIM, 1951–75), 1:99–100; used with permission.

Fig. 7.2a Févin, *Fors seulement,* mm. 1–13, textless version.

measures of this example he set new imitative and nonimitative overlapping counterpoint that is motivically unrelated to the borrowed superius.

Gombert went a step further in attenuating the traditional large-scale structural role of the cantus firmus by using an entire voice from the model to construct only a section within a movement. In *Missa Fors seulement* he employed this technique, using a version of the tenor from Pipelare's second *Fors seulement* setting. In the Qui tollis section of the Gloria for instance, it appears transposed in the altus voice (see Figures 7.3a and 7.3b).

Gombert set new points of imitation to it that are not motivically derived from the model, in addition to adding nonimitative counterpoint. The rhythm of this borrowed tenor stands in marked contrast to the surrounding new counterpoint. Thus Gombert did not integrate the newly composed counterpoint with the borrowed tenor motivically or rhythmically, but owing to the inherent flexibility and versatility of his style, he was able to weave a web of new counterpoint around this borrowed tenor, causing it to blend into the fabric in an artistic fashion. In several places he added a phrase previously borrowed from the Févin chanson to the cantus firmus as a further unifying device.

Gombert also used Pipelare's tenor in a similar fashion in several other sections of this mass. For instance, he used it as the top voice in the Benedictus. Again, borrowing the tenor in this context is simply a short-term

Fig. 7.2b Gombert, *Missa Fors seulement,* Kyrie/Agnus Dei, mm. 1–15 (cantus firmus technique); transcription after Nicolas Gombert, *Opera omnia,* 11 vols., ed. Joseph Schmidt-Görg, CMM 6 (Rome: AIM, 1951–75), 2:89; used with permission.

Fig. 7.3a Pipelare, *Fors seulement II,* mm. 12–23; transcription after Matthaeus Pipelare, *Opera omnia,* 3 vols., ed. Ronald Cross, CMM 34 (n.p.: AIM, 1966–67), 1:12; used with permission.

organizational device used to carry the mass forward in a coherent fashion by reiterating familiar-sounding material at different points. Clearly Gombert had no problems composing an imitative work while employing old-fashioned linear techniques associated with practices from the past.

All these examples illustrate the linear use of borrowed material in an imitative work; it did not matter whether the model was composed in an imitative fashion, in the new Parisian-type style, or in the old-fashioned layered style. Gombert was able to create a context that could accommodate material borrowed in any fashion from any style of model. As we see, the borrowed melody no longer functions as the single guiding organizational device of

Fig. 7.3b Gombert, *Missa Fors seulement,* Qui tollis, mm. 7–18 (cantus firmus technique); transcription after Nicolas Gombert, *Opera omnia,* 11 vols., ed. Joseph Schmidt-Görg, CMM 6 (Rome: AIM, 1951–75), 2:97; used with permission.

the mass. It has become internalized and is used freely as one of many compositional tools. There are further phases in this internalization process: on occasion the entire melody is modified in some significant fashion, typically by rearranging the phrases or by interpolating new phrases within it. When the borrowed melody is used in one of these ways in these chanson masses, it appears sporadically, and the alterations are kaleidoscopic in nature because the changes are not predictable or part of an obvious schematic plan other than to reinterpret familiar material in a continuously changing manner.

Next I will discuss the technique of block interpolation, borrowing a section of chanson material and inserting new passages throughout it. In the previous discussion of masses based on Parisian-type chansons and models from the previous generation composed in a linear fashion, I pointed out that the composer made an attempt to blend these two disparate styles: Franco-Flemish counterpoint and simple homophonic/homorhythmic textures. Apparently Crecquillon did not consider this important: in some of his masses he blended these two textures, and in others he did not. His disregard for blending styles is strikingly evident in the *Missa Doulce memoire.* Of all his chanson masses, this mass extracts the largest blocks from the model: throughout the mass, slightly varied blocks of chanson material are juxtaposed with newly composed passages. This is interesting because the texture of this model, a Parisian-type chanson, stands in marked contrast to Franco-Flemish counterpoint. Crecquillon even went so far as to emphasize the Parisian element in this mass: not only did he begin movements with a block of chanson material, but he scattered blocks throughout the mass without reshaping them into imitative complexes, as he did in other masses. In several instances Crecquillon borrowed such a large block that one could almost call the practice *contrafactum.* For instance, he borrowed the entire third section of the chanson, varied it slightly, and used it as Kyrie II.

In many sections of the mass, he freely juxtaposed passages of borrowed homophonic material with newly composed passages, some of them imitative and some homorhythmic. The Et in terra pax section of the Gloria illustrates this compositional process (see Figures 7.4a and 7.4b).[7]

In this section Crecquillon wrote in a variety of ways to create an interesting musical design. The following numbers are marked in the example: (1) he borrowed a homophonic block and varied the rhythm to accommodate the text; (2) he composed a new passage based on points of imitation; (3) he twice modified a block of chanson material and embellished the harmony without changing it; (4) he paired newly composed voices, alternating between the upper and lower ones in an imitative section; and (5) he composed a new passage in a homorhythmic style to stress the words "Jesu Christe." The end of a line of text demarcates each of these small sections. On occasion one phrase may dovetail into the next; when this occurs, no more than one syllable overlaps.

Fig. 7.4a Sandrin, *Doulce memoire,* mm. 1–17.

Fig. 7.4b Crecquillon, *Missa Doulce memoire,* Gloria, mm. 1–32 (block interpolation); transcription after Thomas Crecquillon, *Opera omnia,* ed. Barton Hudson, Mary Tiffany Ferer, and Laura Youens, CMM 63 (n.p.: AIM, 1974–), 2:3–4; used with permission.

New
(4)
20
mus ti - - - bi pro - pter ma - gnam glo - ri - am tu -
mus ti - - - bi pro - pter ma - gnam glo - ri - am
mus ti - bi...
mus ti - bi...
25
am.
Do -
tu - am.
Do -
New
Domi-ne De - us.Rex cae-le - stis.De - us Pa-ter o - mni - - - po - tens.Do-
Do - mi-ne De - us. Rex cae - le - stis. De - us Pa - ter o - mni - po - tens. Do-
mm. 14–17
(3)
30
New
(5)
mi - ne Fi - li u - ni - ge - ni - te Je - su Chri - ste.
mi - ne Fi - li u - ni - ge - ni - te Je - su Chri - ste.
mi - ne Fi - li u - ni - ge - ni - te Je - su Chri - ste.
mi - ne Fi - li u - ni - ge - ni - te Je - su Chri - ste.

Fig. 7.4b (*cont.*).

Obviously the idea of continuously juxtaposing these disparate styles presented no aesthetic problem for Crecquillon. Perhaps he did not consider these two styles or textures to be in opposition. After all, he wrote his own chansons in both imitative and polyphonically animated homophonic styles.

The simple, somewhat lucid texture of Crecquillon's *Missa Doulce memoire*, just described, completely vanishes in his *Missa Mort m'a privé*. This five-voice mass is much longer and is composed in a dense, imitative style. He modeled it on one of his own chansons, written in a polyphonically animated homophonic style (see Figures 7.5a and 7.5b).[8]

But most of the homophonic elements found in *Missa Doulce memoire* are not present in *Missa Mort m'a privé*. In this mass Crecquillon has used a homophonic model to compose an imitative work. He accomplished this by employing two basic devices: he reiterated a variety of borrowed melodic fragments periodically throughout the movements, and he refashioned chanson blocks. The most interesting aspect of this piece is how he refashioned the homophonic blocks or partial blocks that he borrowed. For instance, in a passage from the Christe, Crecquillon used the technique of partial scaffolding: he selected a segment of chanson material, fragmented it, rearranged its constituent parts, and wrote points of imitation around its various components, creating an imitative frame (see Figure 7.5b). There are numerous variations of this technique present throughout this mass.

Crecquillon's *Missa D'amours me plains* is based on a predominately imitative chanson by Rogier Pathie. Although the overall texture of this chanson is imitative, it does contain brief homophonic passages; the same applies to the mass. In this mass, Crecquillon borrowed a variety of material, including motives, blocks, and phrases. Also, he borrowed both homorhythmic and imitative blocks, but once again, a homorhythmic block frequently was turned into an imitative complex. Variations of all these types of constructions are common in this mass. Many other techniques, such as reiterating a chanson motive throughout a section or simply borrowing a block and juxtaposing new contrapuntal passages, are used throughout this piece: I will not give examples of these structures because there is nothing innovative about the process.

There is, however one feature of this mass that is innovative and clever. Crecquillon selected a large section of chanson material and continuously rearranged blocks of it, adding very little new counterpoint. In other words, he created a new structure using the technique that I have called varied block reiteration. This technique functions in much the same fashion as motivic reiteration, which is a common device in many chanson masses.

Often when one examines a passage constructed by block reiteration, it is difficult to make any structural sense out of these varied, conflated, repeated, rearranged blocks. How did Crecquillon compose such a section,

Fig. 7.5a Crecquillon, *Mort m'a privé,* mm. 1–10 (borrowed scaffold marked).

Fig. 7.5b Crecquillon, *Missa Mort m'a privé,* Christe, mm. 32–41 (partial scaffolding); transcription after Thomas Crecquillon, *Opera omnia,* ed. Barton Hudson, Mary Tiffany Ferer, and Laura Youens, CMM 63 (n.p.: AIM, 1974–), 3:2; used with permission.

and what, if any, was the thought process responsible for this capricious rearranging? Kyrie II serves as a good place to explore this issue. Most of Kyrie II is based on varied, repeated blocks borrowed from mm. 38–50 of the chanson; only sporadically did he add bits of new counterpoint (see Figures 7.6a and 7.6b).[9]

Fig. 7.6a Pathie, *D'amours me plains,* mm. 37–50.

Fig. 7.6b Crecquillon, *Missa D'amours me plains,* Kyrie, mm. 48–69 (varied block reiteration); transcription after Thomas Crecquillon, *Opera omnia,* ed. Barton Hudson, Mary Tiffany Ferer, and Laura Youens, CMM 63 (n,p,: AIM, 1974–), 1:3; used with permission.

Fig. 7.7 Crecquillon, *Missa D'amours me plains,* Kyrie, mm. 48–69 (borrowed superius/bassus); transcription after Thomas Crecquillon, *Opera omnia,* ed. Barton Hudson, Mary Tiffany Ferer, and Laura Youens, CMM 63 (n.p.: AIM, 1974–), 1:3; used with permission.

It appears that initially he was interested not in borrowing blocks of chanson material but rather in borrowing the melodic superius and the somewhat harmonic bassus. In fact, he extracted material from these two voices first. Then he rearranged the phrases in these two parts to create a more tonal superius/bassus frame (see Figure 7.7). After examining this scaffold, it is easy to see how he added the remaining new and borrowed material. He constructed what appears on the surface to be a randomly selected, illogical, multivoiced structure but is actually a carefully planned two-voiced frame, with mostly borrowed harmonies added.

It should be clear that the composers discussed here did not invent new contrapuntal devices but rather relied on compositional practices from the past that they reinterpreted in a familiar context—the setting of the Mass Ordinary—in order to create a new integrated style. Instead of employing a particular technique to create a large-scale structure, they composed a mutable web of counterpoint governed only by personal harmonic considerations that could accommodate or be accommodated by material taken from the model in any fashion and used at any point in the mass. Many different procedures are used for reworking the fragments from the model, which are either stratified, intertwined, fragmented, or juxtaposed with newly composed counterpoint. The recurrence of familiar-sounding borrowed material creates a sense of large-scale structure and cohesion by its repetitive use, and thus articulates this often seamless polyphonic texture.

In constructing these masses it was common for the composer to modify the rhythm of the borrowed material in whatever fashion best suited the compositional exigencies of the moment, and through this flexibility he was able to shape the work into an evenly flowing musical unit. Consequently, it did not matter if the style of the model was different from the style of the mass, because altering the rhythm could create the desired vertical sonorities and textural design. It also did not matter in what fashion the material was borrowed. Individual compositional preferences and the context in which the borrowed material was placed determined the style.

As a final illustration of the insufficiency of analysis inspired by quantitative criteria such as the amount and kind of borrowing, and the need for analysis based on compositional process, contextual design, and personal stylistic preferences, consider Figure 7.8. The opening measures of borrowed material in the Gloria of Gombert's mass and the corresponding section from Palestrina's mass are based on the same model, Cadéac's *Je suis desheritée* (see Figure 7.1a). As one would expect, both composers borrow material at the beginning of the Gloria. If one looks only at what is borrowed, it is not obvious who composed these scaffolds. Gombert simply borrows the superius/tenor framework of the chanson with minor variations, and Palestrina creates a new structure based on chanson material. These two types of structural design are common to both composers. But when one examines the newly composed material in each of these passages, it is obvious which is by Gombert and which by Palestrina (see Figure 7.9). Palestrina is interested in setting the text in a clear, rhythmic fashion, while Gombert has no such restriction. Palestrina also develops a countertheme, to which he seems to give as much weight as the main theme. This makes for a balanced, symmetrical form, so unlike Gombert, who prefers free counterpoint.

If one listens to these two masses carefully, it becomes apparent that Palestrina was familiar with Gombert's work. There are passages in Palestrina's mass that are reminiscent of ones in Gombert's mass. Figure 7.10 shows the two borrowed scaffolds from the opening of the Qui tollis. Again, based only on what is borrowed, it is not clear which one is by which composer. Palestrina embellishes the borrowed theme in the tenor (see Figure 7.10b, mm. 46–47). This embellishment is a variant of a newly composed motive by Gombert (see Figure 7.11a) that Palestrina uses again, as shown in Figure 7.11b.

Fig. 7.8 (a) Gombert, *Missa Je suis desheritée,* Gloria, mm.1–6 (borrowed material only); transcription after Nicolas Gombert, *Opera omnia,* 11 vols., ed. Joseph Schmidt-Görg, CMM 6 (Rome: AIM, 1951–75), 1:83; used with permission; (b) Palestrina, *Missa Je suis desheritée,* [Missa sine nomine], Gloria, mm. 1–5 (borrowed material only); transcription after Giovanni Pierluigi da Palestrina, *Le opere complete,* 35 vols., ed. Raffaele Casimiri et al. (Rome: Istituto Italiano per la storia della musica, Edizione Scalera, 1939–99), 21:53; used with permission.

Once more, if one examines the music that is not borrowed in these two passages, it is very clear which passage is by Gombert and which by Palestrina (see Figures 7.11a and 7.11b). Palestrina is sensitive to text declamation—illustrated by his small rhythmic modifications to accommodate accent—whereas Gombert is content to retain the original rhythm, adding the text without specific attention to this matter.

Fig. 7.9 (a) Gombert, *Missa Je suis desheritée,* Gloria, mm. 1–6 (new counterpoint); transcription after Nicolas Gombert, *Opera omnia,* 11 vols., ed. Joseph Schmidt-Görg, CMM 6 (Rome: AIM, 1951–75), 1:83; used with permission; (b) Palestrina, *Missa Je suis desheritée* [Missa sine nomine], Gloria, mm. 1–5 (new counterpoint); transcription after Giovanni Pierluigi da Palestrina, *Le opere complete,* 35 vols., ed. Raffaele Casimiri et al. (Rome: Istituto Italiano per la storia della musica, Edizione Scalera, 1939–99), 21:53; used with permission.

As for structure, the dense, jagged overlapping imitative and nonimitative phrases and the lack of a clearly articulated imitative design point to Gombert as the composer of the first scaffold. The second scaffold, with its lucid imitative structure, is typical of Palestrina's style. A closer look at these same two newly composed structures shows that Palestrina used parts of Gombert's mass as a model for his own. Notice the similarity between the two composers' bass lines at Miserere nobis, and between corresponding altus phrases in the same complex (Gombert, mm. 75–77, and Palestrina, mm. 46–49). In his usual style, Palestrina creates a much tighter, more compact imitative structure by composing the altus and bassus in imitation, with the tenor picking up the second half of the phrase. What is significant

Fig. 7.10 (a) Gombert, *Missa Je suis desheritée,* Qui tollis, mm. 72–76 (borrowed material only); transcription after Nicolas Gombert, *Opera omnia,* 11 vols., ed. Joseph Schmidt-Görg, CMM 6 (Rome: AIM, 1951–75), 1:86; used with permission; (b) Palestrina, *Missa Je suis desheritée* [Missa sine nomine], Qui tollis, mm. 44–52 (borrowed material only); transcription after Giovanni Pierluigi da Palestrina, *Le opere complete,* 35 vols., ed. Raffaele Casimiri et al. (Rome: Istituto Italiano per la storia della musica, Edizione Scalera, 1939–99), 21:55–56; used with permission.

Fig. 7.11a (a) Gombert, *Missa Je suis desheritée,* Qui tollis, mm. 72–79 (new counterpoint); transcription after Nicolas Gombert, *Opera omnia,* 11 vols., ed. Joseph Schmidt-Görg, CMM 6 (Rome: AIM, 1951–75), 1:86; used with permission.

is that the borrowed bassus passage does not imitate anything. Is this a little bit of the Gombert influence, a nonimitative phrase?

To see that the previous example is not an isolated incident, consider the following example (see Figure 7.12). Figure 7.12b is simply a variant of 7.12a, but it appears obvious that Palestrina, whether consciously or not, borrowed ideas from Gombert. After all, the variation is a very melodic, memorable phrase. Once again, notice Palestrina's articulate placement of

Fig. 7.11b Palestrina, *Missa Je suis desheritée* [Missa sine nomine], Qui tollis, mm. 44–52 (new counterpoint); transcription after Giovanni Pierluigi da Palestrina, *Le opere complete,* 35 vols., ed. Raffaele Casimiri et al. (Rome: Istituto Italiano per la storia della musica, Edizione, Scalera, 1939–99), 21:55–56; used with permission.

this material: the superius/tenor frame and the altus/bassus frame. Gombert presents his material in an asymmetric, less schematic fashion.

These examples suggest that the next phase in the study of borrowing procedures should focus on what is not borrowed from the model and the role of the text in determining stylistic decisions. For instance, it is precisely Palestrina's squaring off of rhythms and dissonance control to accommodate the text accentuation that separate him from Gombert. After all, the form is driven by the text, and the text is what keeps us moving forward in an informed fashion when the familiar music material drops

Fig. 7.12 (a) Gombert, *Missa Je suis desheritée,* Credo, mm. 185–92 (borrowed motive and variation); transcription after Nicolas Gombert, *Opera omnia,* 11 vols., ed. Joseph Schmidt-Görg, CMM 6 (Rome: AIM, 1951–75), 1:94; used with permission; (b) Palestrina, *Missa Je suis desheritée* [Missa sine nomine], Gloria, mm. 70–77 (borrowed motive and variation); transcription after Giovanni Pierluigi da Palestrina, *Le opere complete,* 35 vols., ed. Raffaele Casimiri et al. (Rome: Istituto Italiano per la storia della musica, Edizione Scalera, 1939–99), 21:57; used with permission.

out, creating breaks in our musical labyrinth of borrowed chanson passages and phrases.

By codifying the diverse compositional techniques used in modeling masses on chansons during this period, one can begin to ask more interesting questions. For example, what characterizes an individual's style and how does the same borrowed passage become transformed into a new style when placed in new contextual situation? Perhaps Zarlino's statement on the importance of variation, which he reiterates in various ways in Part III of *Le istitutioni harmoniche* (1558), is a helpful guide for understanding such endeavors.

> [I]t is not only permitted but admirable to duplicate a passage or melody as many times as one wishes if the counterpoint is always different and varied. For such repetitions strike us as being somehow ingenious, and we should try to write them wherever they seem suitable; they are good and not faulty. To an intelligent person they are signs of a lively spirit and an abundance of invention.[10]

The mid-sixteenth-century chanson mass might be the paradigm of this ideal.

Notes

A shorter version of this paper was read at the annual meeting of the American Musicological Society in Pittsburgh, 1994. I would like to thank Bonnie J. Blackburn for her thoughtful comments and suggestions.

1. Infrequent references have been made to parody as early as the fourteenth century; see Leo Schrade, "A Fourteenth Century Parody Mass," *Acta Musicologica* 27 (1955): 13–39. For different conceptions of the subject, see Lewis Lockwood, "On 'Parody' as Term and Concept in 16th-Century Music," in Jan La Rue et al., eds., *Aspects of Medieval and Renaissance Music: A Birthday Offering to Gustave Reese* (New York: W. W. Norton, 1966), 560–75; J. Peter Burkholder, "Johannes Martini and the Imitation Mass of the Late Fifteenth Century," *JAMS* 38 (1985): 470–523; Burkholder, "Communication," *JAMS* 40 (1987): 134–39; Leeman L. Perkins, "Communication," *JAMS* 40 (1987): 130–34; and Reinhard Strohm, "Communication," *JAMS* 40 (1987): 576–79. For a detailed discussion and bibliography on musical borrowings, see J. Peter Burkholder, "Borrowing," *New Grove II.*
2. For a description of the various stages leading to the mature parody mass, see Ludwig Finscher, "Loyset Compère and His Works, IV: The Masses and Mass Sections," *Musica Disciplina* 13 (1959): 123–54.
3. For a discussion of the influence of the style of the model, see Lewis Lockwood, "A View of the Early Sixteenth-Century Parody Mass," in Albert Mell, ed., *Queens College of the City University of New York, Department of Music: Twenty-Fifth Anniversary Festschrift (1937–1962)* (Flushing, N.Y.: Queens College Press, 1964), 53–78.
4. For an in-depth discussion of this issue, see Cathy Ann Elias, "Imitation, Fragmentation, and Assimilation of Chansons in the Masses of Gombert, Clemens, and Crecquillon: A Kaleidoscopic Process" (Ph.D. dissertation, University of Chicago, 1994).
5. For doubts concerning the authorship of this mass, see Eric Jas, "Nicolas Gombert's *Missa Fors seulement*: A Conflicting Attribution," *Revue belge de musicologie/Belgisch tijdschrift voor muziekwetenschap* 46 (1992): 163–77. This essay is not the appropriate place to argue authorship, but if one concludes that this mass is by another composer, it only strengthens the evidence supporting the shift in compositional process described herein.
6. The Févin example is transcribed from 1538^{9}; for a modern edition of the Cadéac chanson, see Robert Eitner, ed., *Sechzig Chansons zu vier Stimmen aus der ersten Hälfte des 16. Jahrhunderts von französischen und niederländischen Meistern*, Publikationen älterer praktischer und theoretischer Musikwerke 23 (Leipzig: Breitkopf & Härtel, 1899), 20–21.
7. The Sandrin chanson is transcribed from 1544^{10}.
8. The chanson is transcribed from 1545^{14}.
9. The chanson is transcribed from 1543^{16}.
10. Gioseffo Zarlino, *The Art of Counterpoint: Part Three of Le istitutioni harmoniche, 1558*, trans. Guy A. Marco and Claude V. Palisca (New Haven and London: Yale University Press, 1968), 154.

8
Melodic Citation in the Sixteenth-Century Motet

MICHELE FROMSON

In an article published in the *Journal of the American Musicological Society* in 1994, I presented a controversial new hypothesis to explain the historical and expressive significance of Adrian Willaert's *Musica nova* (1559).[1] I proposed that this monumental collection, which most Renaissance scholars have associated with the politics and culture of mid-sixteenth-century Venice, may instead have been compiled to commemorate the republican political traditions of Willaert's Florentine patrons, many of whom were living as expatriates in Venice at that time. This hypothesis drew on documentary, historical, literary, religious, and more broadly cultural evidence, most of which is well known. What was new about my argument, and what pointed to a Florentine context for *Musica nova,* was my claim that Willaert embedded short citations of liturgical chant in his dense polyphony.

This musical evidence and, perhaps more importantly, the historical conclusions I drew from it have raised eyebrows among many Renaissance musicologists. Some have questioned whether chant citations exist at all. Might it not have been the case, they counter, that melodic lines whose pitches conform to a liturgical melody are simply the unintended by-product of a polyphonic idiom based on the general style of chant? Others have posed a more general question: even if we accept that Willaert sometimes cited liturgical chant, can we identify these citations with sufficient objectivity and consistency to support an entirely new historical interpretation of *Musica nova?*

The purpose of the present essay is to address these questions.[2] I will not try to develop a comprehensive theory of melodic citation, for that would surely be impossible with such a vast repertory. Instead, by closely examining representative motets from the period, I shall explain how to identify chant citations and demonstrate how uncommon they are in motets of the mid- to late sixteenth century. I shall also describe the range of expressive purposes they usually serve, and develop basic musical and extramusical criteria for dealing with ambiguous cases. Before closing I shall consider the general methodological issues that the study of chant citation raises.

Defining Chant Citations

Presented in Figure 8.1a is a passage I believe contains a straightforward chant citation. Here, at the beginning of Verdelot's motet *In te Domine speravi*, we find the first phrase of the well-known hymn *Te Deum laudamus* (Figure 8.1c).[3] Significantly, Verdelot did not paraphrase the chant setting of his motet text (Figure 8.1d),[4] as was commonly done in the early sixteenth century.[5] In this and subsequent examples, notes drawn from the chant are marked with *x*'s.

As a comparison of Figures 8.1b–d and 8.1e will demonstrate, the chant in Verdelot's motet has been manipulated much as Palestrina would do several decades later in his paraphrase mass *Missa Te Deum*, whose Sanctus is given in Figure 8.1e.[6] Both passages draw attention to the chant in the highest voice at the very beginning of the polyphony. All the notes of the first phrase are presented in new mensural notation, with note repetitions changed to accommodate the sung texts. There is one important difference between Palestrina's paraphrase and Verdelot's citation, however. In Verdelot's motet only the first phrase of the chant appears, whereas in Palestrina's mass successive phrases are presented across its five sections, thereby providing many opportunities to recognize the monophonic model.

A second example of chant citation can be seen at the beginning of Willaert's motet *Verbum iniquum et dolosum* (Figure 8.2a; the full contrapuntal texture is reproduced as Figure 8.3, where superimposed dots indicate notes from *Verbum iniquum* and *x*'s signify notes from *Judas mercator*). At first the opening head motive F-G-A-G (bracketed in Figure 8.3) suggests the responsory setting of the sung text (Figure 8.2b);[7] however, as the motive develops, a different responsory is invoked: *Judas mercator pessimus* (Figure 8.2c).[8] This longer motive soon dominates the cantus, quintus, and tenor voices, with the longest statement appearing in the quintus from m. 11ff.

Although only the motet text is sung during a citation, the chant text is implicitly referenced by its original melody. This means that during such passages, the motet is implicitly polytextual, much like contemporaneous

Fig. 8.1 (a) Verdelot, *In te Domine speravi,* opening imitative segment, cantus and altus, mm. 1–7; (b) Verdelot, *In te Domine speravi,* cantus (transposed down a perfect fourth), mm. 1–7; (c) *Te Deum laudamus,* opening phrase (LU 1834, among others); (d) *In te Domine speravi,* opening phrase (LU 1008); (e) Palestrina, *Missa Te Deum laudamus,* Sanctus, cantus, mm. 1–8.

masses built on untexted cantus firmi.[9] And as in those masses, the semantic relationship of the sung to the cited text can be of considerable historical interest to us because of information it can provide about the expressive significance or the historical context of the polyphonic work. This kind of "implicit polytextuality" has been discussed by Margaret Bent in relation to the untexted tenors of fourteenth-century motets, and her observations about that earlier repertory are directly relevant to motets under consideration here:

Fig. 8.2 (a) Willaert, *Verbum iniquum et dolosum,* quintus, mm. 11–20; (b) *Verbum iniquum et dolosum,* opening phrase (transposed down a perfect fifth) (LA 278); (c) *Judas mercator pessimus,* opening phrase (transposed down a major second) (LU 644).

> Even if the portion of chant adopted for a motet tenor lacks text in the manuscripts that have come to us, for the composer who selected it, it would have been inseparable from its words. It is particularly relevant to remember that the biblical or liturgical text from which a tenor is drawn invites us to invoke further levels of unstated, or at least understated, context. Indeed, the full verbal context of a chant must often have prompted its choice, as cases of clear symbolic or intertextual significance attest.[10]

The same holds true for citations: the text to which the borrowed melody alludes will usually have some semantic relationship to the words being sung at that moment in the motet. Often one finds a shared subject matter that may suggest further avenues of investigation. Given the goals and scope of the present essay, however, my explications of these interrelationships will necessarily be limited to an identification of the most obvious semantic connections between sung and cited texts.[11]

In the case of Verdelot's *In te Domine speravi,* the texts of both the motet and the cited hymn address the Lord directly and affirm their faith in Him. The Lord is also extolled, perhaps not coincidentally, in the instructional rubric that appears in the quintus voice of at least one manuscript source for the work, the Newberry/Oscott Partbooks.[12]

Motet: In te Domine speravi
  In thee, Lord, do I put my trust
Cited chant: Te Deum laudamus
  We praise you, O Lord
Canonic rubric: Exaltata est magnificentia tua super celos
  Thy magnificence is elevated above the heavens.

• = notes from LA278 (Verbum iniquum)

x = notes from LU644 (Judas mercatar)

Fig. 8.3 Willaert, *Verbum iniquum–Duo rogari,* opening segment, mm. 1–25; transcription after Adrian Willaert, *Opera omnia,* ed. Hermann Zenck et al., CMM 3 (Rome: AIM, 1950–), 3:1–2; used with permission.

In Willaert's *Verbum iniquum et dolosum* the sung and cited texts are equally complementary. The opening phrase of the motet, "Lying words and deceit," names two evils personified by the biblical figure being castigated in the chant: Judas, the "wretched trader [who] . . . for a handful of coins . . . delivered Christ to the Jews."

> Motet: Verbum iniquum et dolosum
> Lying words and deceit put far from me, O Lord.
> Cited chant: Judas mercator pessimus
> Judas, a wretched trader, betrayed the Lord with a kiss. The Lord, an innocent lamb, did not refuse the kiss from Judas. For a handful of coins he delivered Christ to the Jews.

When a composer such as Verdelot or Willaert cited a chant, his reference is usually prominently displayed in the new polyphony, appearing at the beginning of a composition (Figures 8.1a and 8.3), the beginning of a major section, or less often, the end of the work. In imitative textures the chant normally generates the leading motive, while in nonimitative counterpoint it occupies a structural contrapuntal voice. Citations are carefully coordinated with the phrase of text being sung in that polyphonic segment. They normally span a grammatically self-sufficient phrase of the motet text. A complete unit of the borrowed melody, however, need not be present for reasons that are easily deduced. The citation need only be of sufficient length to be identifiable, perhaps by an educated listener but more likely by a performer, another composer, or someone studying the partbooks or a handwritten score.[13]

Generally speaking, I would argue that in order for a chant to be recognized as a citation, at least nine consecutive notes must be presented in sequence.[14] Fewer notes tend to result in unclear or multiple references. Occasionally a contrapuntal passage will invoke several different melodies or a single melody to which several texts could be sung (see Figure 8.12 below). If so, then we may suspect that a citation exists but will not be able to pin down the reference, at least not on a purely musical basis.

As a rule, citations refer to familiar chants that were sung frequently during the liturgical year (e.g., *Te Deum laudamus* or *Salve Regina*) or heard on important celebrations such as Christmas or Easter. Sometimes the cited chant will have an extremely distinctive contour, as with *Descendi in hortum nucum* and *Salve Regina* (both are discussed below). In the case of *Salve Regina,* even a short quotation would be recognizable to many modern listeners. For the chant to be identifiable, the citation must be very literal at the beginning with elaboration limited to brief cadential flourishes at phrase endings.[15] Again, the more familiar or distinctive the melody, the more embellishment it can tolerate without losing its identity.

Although most citations involve chant melodies, secular tunes are sometimes found. For instance, in Willaert's *Confitebor tibi Domine,* the third motet of *Musica nova,* there is a memorable phrase from Bartolomeo Tromboncino's well-known frottola *Ostinato vo' seguire* (Figure 8.4; shared musical material is indicated by a bracket).[16] In this case, the melody serves a more general expressive purpose than we have encountered thus far. Introduced for the phrase "[And all the kings of the earth] shall sing of the ways of the Lord," it illustrates the verb *cantent* (they shall sing) by invoking a well-known and especially beautiful frottola. Tromboncino's music circulated widely in the mid-sixteenth century and could certainly have reached Willaert by way of the Este court in Ferrara, where both composers worked at several points in their careers.[17]

Motet:	
et cantent in viis Domini	And they shall sing of the ways of the Lord
Frottola:	
Ostinato vo'seguire,	Resolutely I shall follow
La magnanima mia impresa.	my noble enterprise.
Fami amor qual voi offesa	Do to me, Love, what you will,
S'io dovesse ben morire	even if I should die.

Secular citations of this kind can be difficult for modern listeners to recognize because the repertory is vast and largely unfamiliar to us today, and we currently have no comprehensive thematic index for it.

Fig. 8.4 (a) Willaert, *Confitebor tibi Domine,* cantus, mm. 119–23 (rebarred); (b) Tromboncino, *Ostinato vo' seguire,* cantus (transposed down a major second), mm. 11–16.

To generalize from the foregoing examples, I would propose that the following characteristics are typical of most chant citations in mid-sixteenth-century motets.

1. The citation is prominently displayed, either at the beginning of the composition or a new contrapuntal segment, in the closing section, as the imitative subject, in the highest-sounding voice, or in a structural contrapuntal voice such as the cantus or tenor.
2. About nine consecutive notes of the chant are presented, with embellishment limited to brief cadential flourishes.
3. The citation spans one statement of a syntactically complete unit of the polyphonic text.
4. If liturgical, the borrowed melody would have been sung regularly during the church year or associated with an important feast.
5. The borrowed melody had wide circulation or, as in the case of Tromboncino's frottola, circulated in regions where the composer worked.
6. A semantic relationship can be seen that connects the words being sung at that moment in the polyphony with those of the cited melody.

The Frequency of Citations in Mid-Sixteenth-Century Motets

Having established these criteria for identifying chant citations, I would like to return to a key question posed at the outset of this essay: whether passages appearing to reference chant might instead be conventional contrapuntal gestures emulating the general style of chant. If so, then citations should be ubiquitous, readily found throughout the contrapuntal texture.

A complete study of this subject would require a much more extensive investigation than can be undertaken in the present context; however, a preliminary, and I would suggest representative, evaluation can be made by sampling motets from the decades around 1540, when much of *Musica nova* was composed. For several reasons, I have selected settings of the text "Congratulamini mihi omnes" (see Figures 8.5 and 8.6). Not only do we have a relatively large number of extant settings but they were written by a variety of composers from different regions in Europe. Furthermore, the group includes three settings by Willaert as well as several others that appeared alongside his in Italian sources of the late 1530s and 40s.[18]

The text "Congratulamini mihi omnes" has two possible continuations: "quia quem quaerebam apparuit mihi," which is appropriate for Easter, and "quia cum essem parvula," which is associated with Marian feasts. Each text

Composer	Approximate Date	Source	Number of Voices
Le Brung	1518	BolC Q19	4
Costanzo Festa	1520s or 1530s	BolC Q20	4
Willaert à 4	late 1520s	BolC Q20	4
Le Heurteur	1534	1534^5	4
Manchicourt	1539	RISM M269	5
Willaert à 5	1539	RISM W1110	5
Clemens	1547	1547^5	4
Crecquillon	1554	1554^3	5
Lassus	1566	RISM L795	6
Massaino	1580	RISM M1270	5
Andrea Gabrieli	1587	RISM G58	8

Fig. 8.5 Settings of "Congratulamini mihi omnes . . . quia quem quaerebam."

also has its own liturgical melody, as shown in Figure 8.7, but those chants are only rarely paraphrased in motet settings of the text.[19]

Close analysis of every voice in the nineteen settings of Figures 8.5 and 8.6 reveals remarkably few citations. Shown in Figures 8.8 and 8.9 are the opening themes from fifteen settings, where one would expect citations to be introduced (by theme I mean the leading motive in an imitative texture or the highest sounding voice in non-imitative counterpoint).[20] *X*'s above each theme mark the longest sequence of notes that fits any chant listed in Bryden and Hughes.[21] According to the criteria developed earlier, citations do not appear in any of the themes in Figure 8.8. In Festa's setting, only four notes follow any known chant; in Willaert's motet, only five notes follow a

Composer	Approximate Date	Source	Number of Voices
Rousée	1534	1534^9	4
Sermisy	1539	1539^{13}	4
Lasson	1539	1539^{13}	4
Willaert	1539	RISM W1108	4
Palestrina	1563	RISM P689	4
Victoria	1572	RISM V1421	6
Costanzo Porta	1580	RISM P5181	8
Donato	1598	RISM D3416	5

Fig. 8.6 Settings of "Congratulamini mihi omnes . . . quia cum essem parvula."

Fig. 8.7 Chant settings of (a) *Congratulamini mihi omnes . . . quia quem quaerebam* (LR 91); (b) *Congratulamini mihi omnes . . . quia cum essem parvula* (LR 248).

chant; in Le Heurteur's composition, there are only four; in Manchicourt's, seven; in Crecquillon's, five; and so forth. Nor are citations found in Figure 8.9: in Lasson's motet only seven notes follow any chant; in Rousee's, only five; and so forth.

Four motets have been omitted from these tables. One is Willaert's five-voice setting of "Congratulamini mihi . . . quia quem quaerebam," which paraphrases the responsory setting of that text throughout (see Figure 8.10).[22] The second is Le Brung's setting, whose cantus voice matches the responsory *Pater peccavi* for eight notes (see Figure 8.11).[23] This "citation" seems questionable for two reasons: not only is the chant absent from the last third of Le Brung's phrase, but the sung and cited texts have little in common. The motet text declares "Rejoice with me all you who love the Lord" whereas the chant expresses the opposite sentiment: "Father I have sinned against heaven and before Thee. I am not worthy to be called Thy son: make me as one of Thy hired servants."

The status of Clemens's opening is equally dubious (see Figure 8.12).[24] Although the first nine notes of its cantus resemble two hymn melodies—*Aeterna Christi munera* and *Iam lucis orto sidere*—the rest of its descending phrase diverges from both ascending hymns. Such discrepancies are not characteristic of chant citations. A different problem arises in Massaino's opening section (see Figure 8. 13).[25] The theme follows the hymn *Apostolorum splendido* for eight notes, but there seems to be no discernible relationship between the sung and cited texts.

In the continuations of our nineteen motets, citations are even less common. In fact, only three might be said to invoke chant melodies at all: the setting by Festa and the two remaining motets of Willaert. Festa cited the Marian antiphon *Descendi, in hortum nucum* for the phrase "Angelus Domini descendit" (see Figure 8.14).[26] Not only did he imitate the

Fig. 8.8 Opening themes from settings of "Congratulamini mihi omnes . . . quia quem quaerebam."

Fig. 8.9 Opening themes from settings of "Congratulamini mihi omnes . . . quia cum essem parvula."

Fig. 8.10 (a) Willaert, *Congratulamini mihi omnes . . . quia quem quaerebam* (5vv.), cantus, mm. 5–14; (b) *Congratulamini mihi omnes . . . quia quem quaerebam* (LR 91).

unusual contour of this chant (Figure 8.14b), which *ascends* a full octave for the phrase "descendi in hortum" before descending a sixth, but the two texts have striking parallels. The motet tells of the angel's descent to Mary to announce the Resurrection, while the chant describes the speaker's descent into the (spiritual) garden to see whether it had blossomed and borne fruit. The citation is further emphasized by being placed immediately after a rousing alleluia refrain whose lower voices imitate chiming church bells (Figure 8.14c). Because this refrain articulates the midpoint (and later the conclusion) of the motet, the citation occurs precisely where the *secunda pars* would begin in a bipartite motet.

Presented in Willaert's four-voice setting of *Congratulamini mihi . . . quia quem quaerebam* is an especially provocative chant: the antiphon *Ecce quam bonum*, whose opening phrase underpins the imitative section with which the *prima* and *secunda pars* of the motet conclude (see Figures 8.15 and 8.16).[27] At first glance, the chant and motet texts seem to have little in common: the singers declare "I saw my Lord" ("Vidi Dominum meum") whereas the chant states "Behold how good and how pleasant it is for brothers to live

Fig. 8.11 (a) Le Brung, *Congratulamini mihi omnes . . . quia quem quaerebam*, cantus, mm. 1–12; (b) *Pater peccavi* (LR 407).

Fig. 8.12 (a) Clemens, *Congratulamini mihi omnes . . . quia quem quaerebam,* cantus, mm. 1–7; (b) *Aeterna Christi* (transposed up a perfect fourth) (ST 140); (c) *Iam lucis orto sidere* (transposed up a perfect fourth) (ST 72).

together in unity." As explained in more detail in my original article about *Musica nova,* however, I believe that this antiphon was cited on account of the powerful Savonarolan overtones of its text, which the controversial friar had set as a *lauda* in the late 1400s.[28] The appearance of this chant during the phrase "Vidi Dominum meum" (I saw my Lord) suggests that someone intimately connected with the motet—a patron, members of Willaert's audience, his singers, or perhaps the composer himself—sympathized with the Savonarolan reform movement.[29]

The foregoing survey of *Congratulamini mihi omnes* suggests that citations are an extremely rare phenomenon in motets of the mid- to late

Fig. 8.13 (a) Massaino, *Congratulamini mihi omnes . . . quia quem quaerebam,* cantus, mm. 1–5; (b) *Apostolorum splendido* (transposed up an octave) (ST 292).

Fig. 8.14 (a) Festa, *Congratulamini mihi omnes . . . quia quem quaerebam,* cantus, mm. 55–62; (b) *Descendi in hortum* (transposed down a major second) (PM 243); (c) Festa, *Congratulamini mihi omnes . . . quia quem quaerebam,* Alleluia refrain, mm. 49–58.

Fig. 8.15 (a) Willaert, *Congratulamini mihi omnes . . . quia quem quaerebam* (4vv.), tenor, mm. 58–62; (b) *Ecce quam bonum* (transposed up a perfect fifth) (LU 295).

sixteenth century. Out of nearly twenty settings, we have found only three clear-cut cases: two by Willaert and one by Festa. Even if the ambiguous openings by Le Brung, Clemens, and Massaino are included, there are only six citations in hundreds of measures of counterpoint, thousands if individual lines are counted. The implications of this evidence are quite clear. Citations are not the inadvertent by-products of a chant-based contrapuntal idiom but rather deliberate references to sacred melodies that were the musical lingua franca of the time.

Other Analytic and Interpretive Issues

Having developed a general profile of chant citations and sampled their frequency in motets of the mid-sixteenth century, I would like to consider some of the more subtle analytic and interpretive questions that are raised by citations. An effective way to address these issues is to compare different citations of a single well-known chant: the Marian antiphon *Salve Regina mater misericordia.*

To date, I have located eighteen possible citations of this well-known melody, as presented in Figures 8.17 and 8.18.[30] In Figure 8.17, none of the passages meets the criteria established above for a citation. Jacquet's melody is too short and its descent from A to D too widely separated by rests to invoke the chant. The passages by Phinot, Manchicourt, Willaert, Crecquillon, and Zarlino simply do not resemble the chant melody for a sufficient number of notes. In Figure 8.18, on the other hand, I would classify passages (d) through (o) as true citations (Figures 8.18b and 8.18c are representative passages from Palestrina's mass on *Salve Regina,* presented to show how similar citations are to standard paraphrase techniques of the period).

Fig. 8.16 Willaert, *Congratulamini mihi omnes . . . quia quem quaerebam* (4vv.), mm. 56–76 (individual statements of the theme are bracketed); transcription after Adrian Willaert, *Opera omnia,* ed. Hermann Zenck et al., CMM 3 (Rome: AIM, 1950–), 1:27–28; used with permission.

Fig. 8.17 Improbable citations of *Salve Regina*.

Fig. 8.18 Likely citations of *Salve Regina*.

(i) Morales, *Andreas Christi famulus*, mm. 77–82

(j) Guerrero, *Ave Virgo sanctissima*, superius 1 & 2, mm. 28–35
(transposed down a major third)

(k) Willaert, *Germinavit radix*, altus, mm. 302–7

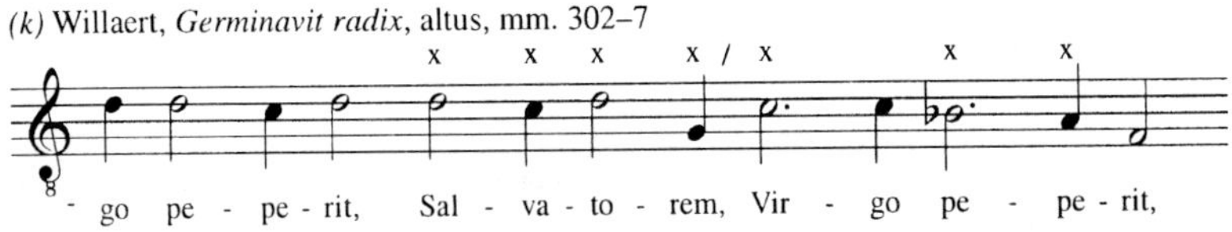

(l) Palestrina, *Nativitas tua Dei genitrix Virgo*, cantus, mm. 2–8

Fig. 8.18 *(cont.)*.

(m) Gombert, *Sancta Maria, succure miseris*, quintus, mm. 5–13
(transposed down a perfect fifth)

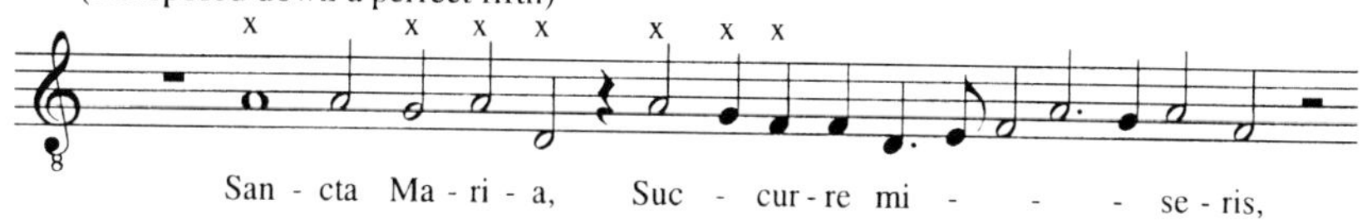

(n) Layolle, *Domine, exaudi orationem meam*, highest voice, mm. 111–17
(transposed down a perfect fourth)

(o) Layolle, *Domine, exaudi orationem meam*, highest voice, mm. 182–87
(transposed down a perfect fourth)

Fig. 8.18 *(cont.)*.

Passages (d) through (h) invoke the antiphon for setting other "Salve" texts: Piéton, *Salve crux digna;* L'Héritier and Crecquillon, *Salve mater salvatoris;* Orazio Vecchi, *Salve radix sancta;* and Viadana, *Salve sancte pater.* Morales (i) and Guerrero (j) do so as well but in the middle of their motets, not at the opening.[31] Both citations appear in a structural voice—the highest voice of Morales and the canonic duo of Guerrero. The frequency with which the antiphon was invoked for "Salve" passages suggests that a chant-based convention may have existed during this period for setting such texts.[32]

The citation in Willaert's *Germinavit radix* (k) functions somewhat differently, with the antiphon invoked for "Salvatorem" (the Savior) rather than "Salve" (Hail). Here, the antiphon refers to the Virgin Mary, whom Willaert's text extols for having borne the Savior, and also draws attention to the assonant relationship of the two texts, much like a musical pun.

> [Virgo peperit] Salvatorem
> [The Virgin has brought forth] a Savior

The citations at the beginning of Palestrina's *Nativitas tua Dei genitrix* (l) and Gombert's *Sancta Maria succure miseris* (m), by contrast, have neither a verbal nor a sonorous connection with the Marian antiphon. Instead they invoke the Virgin Mary, the subject of both motets, as the comfort, hope, and advocate for humankind (see text below).

Palestrina: Nativitas tua, Dei genitrix Virgo
Your birth, O virgin mother of God
Gombert: Sancta Maria, succure miseris
Holy Mary, help those in misery
Antiphon text:
Hail, queen and mother of mercy. Hail our life, comfort and hope. Exiled sons of Eve, with loud voice we call upon you. As we journey in sorrow and lament through this Valley of Tears, we sigh and long for your help. Come then, our advocate, and turn those eyes of pity towards us now. When this time of exile is past, show us Jesus, the blessed fruit of your womb, gentle, loving, and kind Virgin Mary.

In Layolle's motet *Domine, exaudi orationem meam*, each citation serves yet another purpose (Figures 8.18n-o). In the first, which is nearly as long as those of Palestrina and Gombert but displays somewhat more elaboration, the words of the antiphon are only loosely related to those of the motet:

Layolle: Auditam fac mihi mane misericordiam tuam,
Cause me to hear Thy mercy in the morning;

The expressive function here seems to be to associate the Lord's mercy with that of His merciful advocate, Mary. Layolle's second citation, occurring near the end of the motet, functions differently.

Layolle: Gloriam patri et filio, et spiritui sancto.
Glory be to the Father and to the Son, and to the Holy Spirit.

Rather than having a semantic relationship with the text of the lesser doxology, it serves to articulate the formal plan of the motet by referring back to previously heard thematic material. Recalls of this sort are common in this repertory, especially in bipartite motets like Layolle's.

Citations of *Salve Regina*, then, serve a wide range of referential purposes. In Palestrina's mass the antiphon provides the thematic scaffolding for a large, multipartite composition. In the motets of Piéton, L'Héritier, Crecquillon, Orazio Vecchi, Viadana, Morales, and Guerrero the citation is both a conventional setting of "Salve" and a reference to the Virgin Mary as comforter and protectress. This latter function is preeminent in the motets by Willaert, Palestrina, and Gombert (none of whose texts contain the word *Salve*), where the melody draws attention to the subject of their own compositions, the Virgin Mary. Finally, in Layolle's motet, the first citation associates Mary's mercy with that of God, while the second helps establish formal unity in a bipartite composition.

Citations of *Salve Regina* suggest how versatile this compositional device could be. Needless to say, its history remains to be written. Although the phenomenon seems most prevalent in works of the mid-sixteenth century, it has obvious links with cantus firmus procedures of earlier eras. Moreover, as with cantus firmus and paraphrase masses from the fifteenth century or imitative (parodic) masses that proliferated after 1500, the borrowed material in citations is usually prominently displayed, placed in the highest or a structural polyphonic voice, reiterated imitatively or canonically throughout the texture, or located in the opening or closing section of the piece. And as with other polytextual compositional devices (for example, *soggetti cavati* or chant paraphrase in masses), the interrelationship of the cited and sung texts can be of great interest because of information it can provide about the historical context of the polyphonic work.

As a contrapuntal device, citation is most similar to paraphrase, which is common in masses and motets of the same era, for both techniques entail building a new polyphonic fabric around a preexisting, lightly embellished melody. Admittedly, significant differences exist. In a paraphrase motet (though not a mass), the text of the borrowed melody is actually heard in the polyphony, occasionally along with a second set of words sung in a lower voice. This differs from citation, where the text of the borrowed melody is never heard. In citations, the melodic reference would be comprehended only by people who could identify the theme by its opening phrase and could remember the original text. Citation was, therefore, intended for cognoscenti—perhaps an astute listener but more likely professional singers and composers well versed in the liturgical repertory and with direct access to the notated polyphony.

Citation differs from paraphrase in another important way. With paraphrase, successive phrases of the chant would be used to generate thematic material for successive contrapuntal segments, thereby giving the listener many opportunities to identify the chant. In a typical citation, by contrast, only one short phrase of the chant is presented and only within a single contrapuntal segment.[33] This difference suggests why citation may be more prevalent in polyphony written from the early 1500s on. By this time a decisive shift had taken place away from the layered compositional procedures we believe were followed in the fifteenth century, which entailed composing individual voices successively according to a hierarchic scheme that gave priority to the tenor and cantus voices. By 1530, recent research suggests, multiple voices were being created more or less simultaneously, apparently on a segment-by-segment basis.[34]

In a pathbreaking study of compositional process in the late fifteenth and sixteenth centuries, Jessie Ann Owens described such procedures as follows:

> [From the nearly complete absence of scores used for composing vocal music] we must conclude that composers could manage the complicated polyphony of the time without using the format that seems indispensable to us. Working on short segments using quasi-score format or on longer segments using separate parts, they were able to sketch and draft the early versions of their music. . . . The process was essentially additive. . . . In many cases it is clear that composers worked segment by segment even when they had a single line—either an existing tune or a newly composed tenor—as their point of departure. Sometimes they worked first with two "essential" voices and added a third, while at other times the grid of essential versus added shifted from phrase to phrase. . . . Composers of vocal music kept track by reading voices one against another, by adding vertical lines to demarcate segments, and by counting tactus.[35]

Because citations were deployed within individual segments rather than across an entire piece, they were well suited to these new compositional procedures. Perhaps for a composer such as Willaert, citation presented exciting expressive opportunities by allowing him to associate individual phrases of his own text with specific borrowed melodies and, by extension, with the texts to which they allude.

* * *

The purpose of the foregoing essay has been to establish general criteria for identifying citations, to demonstrate how uncommon they are in a representative group of motets from the mid-sixteenth-century motet, to survey the diverse referential functions they can have, and to address the most serious objections raised by my earlier research on *Musica nova.* It is my hope that with this additional information in hand, scholars working on other polyphonic repertories of the sixteenth century will look for citations in their own areas of research, determining whether citations were used or not, and if so, what kind of referential purposes they might have served.[36]

In truth, chant citations, like many other referential musical devices, are comparatively uninteresting in and of themselves. They acquire meaning only in relation to a specific historical context—a particular geographic region, liturgical tradition, school of composers, cultural center, and so forth. Although citations are discovered primarily through intuitive, aural analyses of the individual lines in a polyphonic composition, their scholarly significance hinges on their relationship to historical contexts known primarily through the study of written documentary evidence. The study of citation (and other referential musical devices) therefore brings together two contrasting methodologies that have often been in opposition to one another in studies of Renaissance music—the first predicated on a fundamentally subjective reading of the music text, and the other grounded more objectively in studies of extant written documents. Indeed, in my earlier work on *Musica nova,* the significance of Willaert's citations began to emerge only in relation to the richly documented historical context that other scholars had already established for the collection through their ingenious, painstaking, and incredibly thorough investigations of extant letters, payment records, biographies, legal documents, papal records, poetry and other literary works, religious tracts, and historical chronicles.[37] It was the interrelationship between those fascinating documents and the melodic citations, which I discovered while singing passages from Willaert's polyphony, that led me to propose a new Florentine reading of *Musica nova.*

Notes

Earlier versions of this paper were delivered at the annual joint meeting of the Northern and Southern California Chapters of the American Musicological Society (Getty Museum, Los Angeles, April 1995) and at the Conference on Medieval and Renaissance Music, University of Southampton (July 1996). I am grateful to many conference participants, especially Bonnie Blackburn and Jeffrey Dean, for their comments on my paper. The final version of this article was written during a 1998 NEH Summer Seminar for College Teachers led by Jessie Ann Owens, whose thoughtful comments on this paper are greatly appreciated.

1. Michele Fromson, "Themes of Exile in Willaert's *Musica nova,*" *JAMS* 48 (1994): 442–87.
2. I am most grateful to Joseph Kerman for encouraging me to address these difficult questions in print.
3. For a modern edition of the motet, see H. Colin Slim, ed., *A Gift of Madrigals and Motets* (Chicago and London: University of Chicago Press, 1972), 2:140–51. For a modern edition of the hymn, see Heinrich Glarean, *Dodecachordon,* trans. Clement A. Miller, Musicological Studies and Documents 6 (n.p.: AIM, 1965), 1:149.
4. For this chant, see *Graduale marciano* (BerlS 40608), fol. 201v, as cited in Giulio Cattin, *Musica e liturgia a San Marco: testi e melodie per la liturgia delle ore dal XII ad XVII secolo, dal graduale tropato del Duecento ai graduali cinquecenteschi* (Venice: Fondazione Levi, 1990–92), 2:393.
5. The standard printed thematic index of the chant repertory is John R. Bryden and David G. Hughes, *An Index of Gregorian Chant,* 2 vols. (Cambridge, Mass.: Harvard University Press, 1969). In the musical examples to follow, each chant will be keyed to its modern source (LA, LR, LU, PM, ST, WA) as identified by the bibliographical system from ibid., 1:xiii–xv. Also cited for individual musical examples are manuscript or printed sources from the pre-Tridentine era.
6. For a modern edition of the mass, see Giovanni Pierluigi da Palestrina, *Le opere complete,* ed. Raffaele Casimiri et al. (Rome: Istituto Italiano per la storia della musica, Edizione Scaldera, 1939–99), 25:172–208.
7. This chant is found in LucBC 601; for a facsimile, see *Le Codex 601 de la Bibliothèque capitulaire de Lucques: Antiphonaire monastique (XIIe siècle),* Paléographie musicale 9 (Tournai: Desclée, Lefebvre, 1906), 278.
8. My thanks to Bonnie Blackburn for pointing out the paraphrase in this motet. The chant is found in *Antiphonariu[m] s[ecundu]m morem sancte Romanae ecclesie* (Venice: Antonio Giunta, 1503–[4]), vol. 1, fol. 169v. Medieval sources present a second liturgical setting of text, WA 169, but it has no musical connection with Willaert's motet.
9. There are many studies of paraphrase and cantus-firmus-based polyphony of the sixteenth century, most of which deal with the textual relationship of the borrowed melody to the new polyphony. These studies include the classic monograph by Edgar H. Sparks, *Cantus Firmus in Mass and Motet 1420–1520* (Berkeley and Los Angeles: University of California Press, 1963); Jacquelyn A. Mattfield, "Some Relationships between Texts and Cantus Firmi

in the Liturgical Motets of Josquin des Pres," *JAMS* 14 (1961): 159–83; Albert Dunning, *Die Staatsmotette, 1480–1555* (Utrecht: A. Oosthoek, 1970); and Willem Elders, "Plainchant in the Motets, Hymns, and Magnificat of Josquin des Prez," in *JosCon*, 523–42.

10. Margaret Bent, "Polyphony of Texts and Music in the Fourteenth-Century Motet: *Tribum que non abhorruit/Quoniam secta latronum/Merito hec patimur* and Its 'Quotations,' " in Dolores Pesce, ed., *Hearing the Motet: Essays on the Motet of the Middle Ages and Renaissance* (New York and Oxford: Oxford University Press, 1997), 82–103, at 83.

11. Recent studies of the expressive significance of polytextuality in earlier repertories include Dolores Pesce, "Beyond Glossing: The Old Made New in *Mout me fu grief/Robin m'aime/Portare*," in Dolores Pesce, ed., *Hearing the Motet: Essays on the Motet of the Middle Ages and Renaissance* (New York and Oxford: Oxford University Press, 1997), 28–51 (thirteenth-century motets); Bent, "Polyphony of Texts and Music" (fourteenth-century motets); and Michael Long, "Symbol and Ritual in Josquin's *Missa Di Dadi*," *JAMS* 42 (1989): 1–22.

12. ChiN M91 and SuttonO 4.

13. On the use of scores in this period, see Jessie Ann Owens, *Composers at Work: The Craft of Musical Composition 1450–1600* (New York and Oxford: Oxford University Press, 1997).

14. The simplest way to find out whether a chant has been cited or not is by consulting Bryden and Hughes, *An Index of Gregorian Chant.* Although most of the chants they index were taken from modern chant books, those melodies usually begin as they did in the sixteenth century, making this index useful for determining exemplars for chant citations. Nevertheless, in order to ensure that the modern incipits conform with earlier practice, medieval and Renaissance sources should also be consulted, as I have done for the examples in this article. At present, several large electronic databases are being compiled of chant repertories not covered by Bryden and Hughes. The most extensive, CANTUS, presents texts without music. A new searchable database that includes musical information is the La Trobe University Medieval Music Database, prepared by John Stinson (http://www.lib.latrobe.edu.au/AudioVisual/Stinson/medmusic.htm). It contains, among other sources, full texts and complete melodies (converted into letter notation and relative-interval notation) for a complete annual cycle of liturgical chant from Dominican sources written 1270–1345. In the future, innovative on-line resources of this kind will likely give greater access to chants from specific locales and chronological periods.

15. In this respect, citations resemble techniques used in paraphrase masses of the same era. For instance, see Robert Marshall, "The Paraphrase Technique of Palestrina in His Masses Based on Hymns," *JAMS* 16 (1963): 347–72, at 356–58.

16. Several scholars have suggested that this passage may be a conventional cadential gesture of the period rather than a citation. Extensive searches for this gesture by myself and others, however, have uncovered no similar passages. For a modern edition of the motet, see Willaert, *Opera omnia,* 5:15–23; for Tromboncino's frottola, see Benvenuto Disertori, *Le frottole per canta e liuto intabulate da F. Bossinensis* (Milan: Ricordi, 1964), 354–56.

17. Indeed, research by William Prizer puts Tromboncino and Willaert together in Ferrara in the early 1500s; see William F. Prizer, "Isabella d'Este and Lucrezia Borgia as Patrons of Music: The Frottola at Mantua and Ferrara," *JAMS* 38 (1985): 1–33; and Prizer, "Games of Venus: Secular Vocal Music in the Late Quattrocento and Early Cinquecento," *Journal of Musicology* 9 (1991): 3–56, at 7–9.

18. I am grateful to Jennifer Thomas, Mary Lewis, Jane Bernstein, and Bonnie Blackburn for making their large motet indices available to me in order to locate parallel settings of this text. For a summary of the information contained in their indices as well as others of the Renaissance motet, see the appendix to my review "The Sixteenth-Century Motet: An Update on Published Catalogues and Indexes in Progress," *Notes* 52 (1995–96): 45–54.

19. The chants are found in WorcC F.160; for a facsimile, see *Le Codex F. 160 de la Bibliothèque de la Cathédrale de Worcester: Antiphonaire monastique (XIIIe siècle),* Paléographie musicale 12 (Tournai: Desclée; Paris: Alphonse Picard, 1922–25), 49 (*Congratulamini mihi omnes . . . quia quem quaerebam*) and 132 (*Congratulamini mihi omnes . . . quia cum essem parvula*).

20. Modern editions may be found in the following: Costanzo Festa, *Opera omnia,* ed. Alexander Main and Albert Seay, CMM 25 (n.p.: AIM, 1962–78), 5:44–48; Willaert, *Opera omnia,* 1:26–30 (*Congratulamini mihi . . . quia quem quaerebam,* 4v), 2:18–23 (*Congratulamini*

mihi . . . quia cum essem parvula); Albert Smijers and Arthur Tillman Merritt, eds., *Treize livres de motets parus chez Pierre Attaingnant en 1534 et 1535* (Paris: L'Oiseau-Lyre, 1934–64), 3:104–08 (Le Heurteur), 7:44–49 (Rousée), 14:32–36 (Manchicourt); Thomas Crecquillon, *Opera omnia,* ed. Barton Hudson, Mary Tiffany Ferer, and Laura Youens, CMM 63 (n.p.: AIM, 1974–), 6:103–14; Orlando di Lasso, *Sämtliche Werke,* ed. Franz Xaver Haberl and Adolf Sandberger (Leipzig: Breitkopf & Härtel, 1894–1926), 13:10–18; Andrea Gabrieli, *Edizione nationale delle opera,* ed. Denis Arnold and David Bryant (Milan: Ricordi, 1988–), 11/1:243–56; Mary S. Lewis, ed., *The Gardane Motet Anthologies,* The Sixteenth-Century Motet 13 (New York: Garland, 1993), 64–63 (Lasson), 123–27 (Sermisy); Palestrina, *Le opere complete,* 3:94–96; Tomás Luis de Victoria, *Opera omnia,* ed. Felipe Pedrell (Leipzig: Breitkopf & Härtel, 1902–13), 1:129–32; Costanzo Porta, *Opera omnia,* ed. Siri Cisilino (Padua: Biblioteca Antoniana, 1964–70), 5:148–52; Baldassare Donato, *Il primo libro de motetti a cinque, a sei, et otto voci novamente composti, & dati in luce,* ed. Richard Sherr, The Sixteenth-Century Motet 30 (New York and London: Garland, 1994), 117–23.

21. Bryden and Hughes, *An Index of Gregorian Chant.*
22. For a modern edition, see Willaert, *Opera omnia,* 3:102–6.
23. For Le Brung's motet, see Richard Sherr, ed., *Selections from Bologna, Civico museo bibliografico musicale, MS Q19: Rusconi Codex,* The Sixteenth-Century Motet 6 (New York: Garland, 1989), 1:151–60 (published as Richafort's); for the chant, see *Le Codex 601 . . . de Lucques,* 145.
24. For a modern edition of the motet, see Jacobus Clemens non Papa, *Opera omnia,* ed. Karel Philippus Bernet Kempers and Chris Maas, CMM 4 (Rome: AIM, 1951–76), 9:35–38. ST 140 is based on a twelfth/thirteenth-century French source, ST 72 on a twelfth-century French source.
25. Massaino's motet comes from RISM M1270. ST 292 is based on a twelfth-century source from Einsiedeln.
26. For the antiphon, see *Le Codex 601 . . . de Lucques,* 458.
27. A contrapuntal analysis of this same passage is found in Joshua Rifkin, "Miracles, Motivicity, and Mannerism: Adrian Willaert's *Videns Dominum flentes sorores Lazari* and Some Aspects of Motet Composition in the 1520s," in Dolores Pesce, ed., *Hearing the Motet: Essays on the Motet of the Middle Ages and Renaissance* (New York and Oxford: Oxford University Press, 1997), 243–64, at 249–52. For the chant, see *Le Codex 601 . . . de Lucques,* 93.
28. Shortly after Savonarola's execution, this *lauda* was banned by the Church but almost immediately thereafter became the rallying cry of the reform movement he spawned. Even though the antiphon setting was, and remains, a standard component of the office liturgy, Willaert's citation is best construed as a reference to Savonarola or his followers. For a more extended discussion of this subject, see Fromson, "Themes of Exile in Willaert's *Musica Nova,*" 454ff. A variant of this antiphon, which appears in some liturgical books of the sixteenth century, was cited in Morales's motet *Hec est vera martyr* (This is the true martyr), found in 1546^{9}. To my knowledge, the significance of this citation has not yet been determined. A fascinating book about Savonarola and music is Patrick Macey, *Bonfire Songs: Savonarola's Musical Legacy* (Oxford: Clarendon, 1998).
29. Such a conclusion is strongly supported by the appearance of Willaert's motet in the Newberry/Oscott partbooks, which Colin Slim convincingly linked to the cult of Savonarola in Florence in the late 1520s. See Slim, *A Gift of Madrigals and Motets,* especially 1:55–65 and 69–74.
30. The *Salve regina* excerpt is taken from VenMC 1602, fol. 146v. For the other works, see Jacquet of Mantua, *Opera omnia,* ed. Philip T. Jackson and George Nugent, CMM 54 (n.p.: AIM, 1971–86), 5:208–11; Dominique Phinot,*Opera omnia,* ed. Janez Höfler and Roger Jacob, CMM 59 (n.p.: AIM, 1972–82), 4:18–29; Willaert, *Opera omnia,* 13:32–36 (*Mentre che'l cor*), 5:42–45 (*Germinavit radix*); Gioseffo Zarlino, *On the Modes: Part Four of Le Istitutioni harmoniche, 1558,* trans. Vered Cohen (New Haven and London: Yale University Press, 1983), 56; Crecquillon, *Opera omnia,* 6:4–9 (*Adiuva nos Deus*) and 9:77–85 (*Salve mater salvatoris*); Palestrina, *Le opere complete,* 28:185–221 (*Missa Salve Regina*) and 3:78–82 (*Nativitas tua Dei genitrix Virgo*); 1539^{10} (Piéton); Johannes Lheritier, *Opera omnia,* ed. Leeman L. Perkins, CMM 48 (n.p.: AIM, 1969), 1/2:300–7; RISM V1006 (Vecchi); RISM V1354 (Viadana); Cristóbal de Morales, *Opera omnia: Motetes I–XXV,* ed. Higinio Anglès, Monumentos de la musica española 13 ([Rome]: Consejo superior de Investigaciones

Cientificas, Delagación de Roma, 1953), 157–65; Francisco Guerrero, *Ave Virgo sanctissima,* ed. Bruno Turner, Renaissance Performing Scores A, Spanish Church Music 28 (London, Mapa mundi, 1979); Nicolas Gombert, *Opera omnia,* ed. Joseph Schmidt-Görg, CMM 6 (Rome: AIM, 1951–75), 9:80–86; Francesco de Layolle, *Masses and Penitential Psalms,* ed. Frank A. D'Accone, CMM 32/6 (n.p.: AIM, 1973), 125–32.

31. I would like to thank Alejandro Planchart and Owen Rees for bringing the Morales and Guerrero citations, respectively, to my attention.
32. Such a hypothesis deserves a more extended investigation. Research I conducted for an earlier study revealed that by 1570 conventions had developed for at least four liturgical motet texts and were widely used by north-Italian composers. See my "Imitation and Innovation in the North-Italian Motet, 1560–1605" (Ph.D. dissertation, University of Pennsylvania, 1988). The four texts investigated there were "Veni sponsa Christe," "Surrexit pastor bonus," "Quem vidistis pastores dicite," and "O sacrum convivium."
33. Although one could argue that the opening phrase of a chant is its most memorable part, it is entirely possible that subsequent phrases were cited. I have not dealt with this possibility primarily because Bryden and Hughes's *An Index of Gregorian Chant* contains only chant incipits. The La Trobe University Medieval Music Database contains entire melodies of fourteenth-century Dominican chant that are fully searchable.
34. Theoretical writings describing this compositional shift are examined in Owens, *Composers at Work,* especially chapters 2 and 7, 169ff.
35. Ibid., 196.
36. A citation of the chant *Asperges me* has been found by Peter Bergquist in the fourth penitential psalm by Lassus. See Orlando di Lasso, *The Seven Penitential Psalms and Laudate Dominum de caelis,* ed. Peter Bergquist, Recent Researches in the Music of the Renaissance 86–87 (Madison, Wisconsin: A-R, 1990), 81–108, at 89 (verse 8, mm. 1–5). Bergquist's analysis is on p. xi. I am grateful to Jessie Ann Owens for calling my attention to Bergquist's discovery.
37. Nearly all these documents are transcribed in Jessie Ann Owens and Richard J. Agee, "La stampa della 'Musica nova' di Willaert," *Rivista italiana di musicologia* 24 (1989): 219–305. See also Ignace Bossuyt, "O socii durate: A Musical Correspondence from the Time of Phillip II," *Early Music* 26 (1998): 432–44.

Bibliography

Alighieri, Dante. *Vita Nuova: Italian Text with Facing English Translations*. Edited by Dino S. Cervigni and Edward Vasta. Notre Dame and London: University of Notre Dame Press, 1995.

Allsen, Michael. "Style and Intertextuality in the Isorhythmic Motet, 1400–1440." Ph.D. dissertation, University of Wisconsin, Madison, 1992.

Ambros, August Wilhelm. *Geschichte der Musik III*. Leipzig: F. E. C. Leuckart, 1868; 3rd ed., revised by Otto Kade, 1893.

Anglès, Higini. *La música a Catalunya fins al segle XIII*. Barcelona: Institut d'estudis Catalans/ Biblioteca de Catalunya, 1935.

Antiphonariu[m] s[ecundu]m morem sancte Romanae ecclesie. Venice: Antonio Giunta, 1503–[4].

Astell, Ann W. *The Song of Songs in the Middle Ages*. Ithaca: Cornell University Press, 1990.

Atlas, Allan. *Renaissance Music: Music in Western Europe, 1400–1600*. New York and London: W. W. Norton, 1998.

Baxandall, Michael. *Giotto and the Orators: Humanist Observers of Painting in Italy and the Discovery of Pictorial Composition, 1350–1450*. Oxford: Clarendon, 1971.

Bent, Margaret. "Deception, Exegesis and Sounding Number in Machaut's Motet 15." *Early Music History* 10 (1991): 15–27.

———. "Polyphony of Texts and Music in the Fourteenth-Century Motet: *Tribum que non abhorruit/Quoniam secta latronum/Merito hec patimur* and Its 'Quotations.'" In Dolores Pesce, ed., *Hearing the Motet: Essays on the Motet of the Middle Ages and Renaissance*, 82–103. New York and Oxford: Oxford University Press, 1997.

———. "*Resfacta* and *Cantare Super Librum*." *JAMS* 36 (1983): 371–91.

———. "Some Criteria for Establishing Relationships between Sources of Late-Medieval Polyphony." In Iain Fenlon, ed., *Music in Medieval and Early Modern Europe: Patronage, Sources and Texts*, 295–318. Cambridge: Cambridge University Press, 1981.

Bergsagel, John. "The Practice of *Cantus Planus Binatum* in Scandanavia in the 12th to 16th Centuries." In Cesare Corsi and Pierluigi Petrobelli, eds., *Le Polifonie primitive in Friuli e in Europa: Atti del congresso internazionale Cividale del Friuli, 22–24 agosto 1980*, 63–82. Rome: Edizione Torre d'Orfeo, 1989.

Bernstein, Lawrence F. "*La Courone et fleur des chansons a troys*: A Mirror of the French Chanson in Italy in the Years between Ottaviano Petrucci and Antonio Gardano." *JAMS* 26 (1973): 1–68.

Berry, Mary [Mother Thomas More]. "The Performance of Plainsong in the Later Middle Ages and the Sixteenth Century." *Proceedings of the Royal Musical Association* 92 (1965–66): 121–34.

Beumer, Johannes. "Die marianische Deutung des Hohen Liedes in der Frühscholastik." *Zeitschrift für Katholische Theologie* 76 (1954): 411–39.

Binchois, Gilles. *The Sacred Music of Gilles Binchois.* Edited by Philip Kaye. Oxford and New York: Oxford University Press, 1992.
Blackburn, Bonnie J. "A Lost Guide to Tinctoris's Teachings Recovered." *Early Music History* 1 (1981): 29–116.
———. "Messages in Miniature: Pictorial Programme and Theological Implications in the Alamire Choirbooks." *Yearbook of the Alamire Foundation* 5 (forthcoming).
———. "Music and Festivities at the Court of Leo X: A Venetian View." *Early Music History* 11 (1992): 1–37.
———. "On Compositional Process in the Fifteenth Century." *JAMS* 40 (1987): 210–84.
Bloxam, M. Jennifer. "Cantus firmus." *New Grove II.*
———. "In Praise of Spurious Saints: The *Missae Floruit egregiis* by Pipelare and La Rue." *JAMS* 44 (1991): 163–220.
———. "On the Origins, Contexts, and Implications of Busnoy's Plainsong Cantus Firmi: Some Preliminary Remarks." In Paula Higgins, ed., *Antoine Busnoys: Method, Meaning, and Context in Late Medieval Music,* 71–88. Oxford: Clarendon, 1999.
———. "Sacred Polyphony and Local Traditions of Liturgy and Plainsong: Reflections on Music by Jacob Obrecht." In Thomas Forrest Kelly, ed., *Plainsong in the Age of Polyphony,* 140–77. Cambridge and New York: Cambridge University Press, 1992.
———. "A Survey of Late Medieval Service Books from the Low Countries: Implications for Sacred Polyphony, 1460–1526." Ph.D. dissertation, Yale University, 1987.
Bossuyt, Ignace. "O socii durate: A Musical Correspondence from the Time of Philip II." *Early Music* 26 (1998): 432–44.
Bowers, Roger, and Andrew Wathey. "New Sources of English Fourteenth- and Fifteenth-Century Polyphony." *Early Music History* 3 (1983): 123–73.
Brauner, Mitchell P. "Traditions in the Repertory of the Papal Choir in the Fifteenth and Sixteenth Centuries." In Richard Sherr, ed., *Papal Music and Musicians in Late Medieval and Renaissance Rome,* 167–74. Oxford: Clarendon, in association with the Library of Congress, Washington, D. C., 1998.
Brobeck, John T. "Musical Patronage in the Royal Chapel of France under Francis I (r. 1515–1547)." *JAMS* 48 (1995): 187–239.
Brown, Howard Mayer. "Emulation, Competition, and Homage: Imitation and Theories of Imitation in the Renaissance." *JAMS* 35 (1982): 1–48.
———. "Improvised Ornamentation in the Fifteenth-Century Chanson." *Quadrivium* 12 (1971): 238–58.
———. "Music and Ritual at Charles the Bold's Court: The Function of Liturgical Music by Busnoys and His Contemporaries." In Paula Higgins, ed., *Antoine Busnoys: Method, Meaning, and Context in Late Medieval Music,* 53–70. Oxford: Clarendon, 1999.
———. "Notes Towards a Definition of Personal Style: Conflicting Attributions and the Six-Part Motets of Josquin and Mouton." In Willem Elders with Frits de Haen, eds., *Proceedings of the International Josquin Symposium, Utrecht 1986,* 185–207. Utrecht: Vereniging voor Nederlandse Muziekgeschiedenis, 1991.
Brownlee, Kevin. "Machaut's Motet 15 and the *Roman de la Rose*: The Literary Context of *Amours qui a les pouoir/Faus Samblant m'a deceü/Vidi Dominum.*" *Early Music History* 10 (1991): 1–14.
Bruck, Robert. *Friedrich der Weise als Förderer der Kunst.* Strasbourg: J. H. E. Heitz, 1903; reprint ed., Nendeln, Liechtenstein: Kraus, 1979.
Bryden, John R., and David Hughes. *An Index of Gregorian Chant.* 2 vols. Cambridge, Mass.: Harvard University Press, 1969.
Burkholder, J. Peter. "Borrowing." *New Grove II.*
———. "Communication." *JAMS* 40 (1987): 134–39.
———. "Johannes Martini and the Imitation Mass of the Late Fifteenth Century." *JAMS* 38 (1985): 470–523.
Burne, Martin Joseph. "Mass Cycles in Early Graduals: A Study of the Ordinary of the Mass Cycles Found in Medieval and Renaissance Graduals in Libraries in the United States." Ph.D. dissertation, New York University, 1956.
Busnoys, Antoine. *Collected Works: The Latin-Texted Works.* 2 vols. Edited by Richard Taruskin. Masters and Monuments of the Renaissance 5. New York: Broude Trust, 1990.

Camille, Michael. *The Gothic Idol: Ideology and Image-Making in Medieval Art.* Cambridge: Cambridge University Press, 1989.

Caron, Firmin. *Opera omnia.* Edited by Jenny Hodgson. CMM 111. Middleton, Wisc.: AIM, forthcoming.

Cattin, Giulio. *Musica e liturgia a San Marco: testi e melodie per la liturgia delle ore dal XII al XVII secolo, dal graduale tropato del Duecento ai graduali cinquecenteschi.* 4 vols. Venice: Fondazione Levi, 1990–92.

Census-Catalogue of Manuscript Sources of Polyphonic Music, 1400–1550. 5 vols. Vol. 1 edited by Charles Hamm and Herbert Kellman. Vols. 2–5 edited by Herbert Kellman. Renaissance Manuscript Studies 1. Neuhausen-Stuttgart: Hänssler/AIM, 1979–88.

Clemens non Papa, Jacobus. *Opera omnia.* 21 vols. Edited by Karel Philippus Bernet Kempers and Chris Maas. CMM 4. Rome: AIM, 1951–76.

Clinkscale, Edward Henry. "The Complete Works of Antoine de Févin." 2 vols. Ph.D. dissertation, New York University, 1965.

Le Codex 601 de la Bibliothèque capitulaire de Lucques: Antiphonaire monastique (XIIe siècle). Paléographie musicale 9. Tournai: Desclée, Lefebvre, 1906.

Le Codex F. 160 de la Bibliothèque de la Cathédrale de Worcester: Antiphonaire monastique (XIIIe siècle). Paléographie musicale 12. Tournai: Desclée; Paris: Alphonse Picard, 1922–25.

Cohen, Judith. "Munus ab ignoto." *Studia musicologica Academiae scientiarum hungaricae* 22 (1980): 187–204.

———, ed. *Six Anonymous L'homme armé Masses in Naples, Biblioteca Nazionale, MS VI E 40.* CMM 85. Neuhausen-Stuttgart: Hänssler/AIM, 1981.

Connolly, James L. *John Gerson: Reformer and Mystic.* Université de Louvain, Recueil de travaux publiés par les membres des Conférences d'histoire et de philologie II/12. Leuven: Librairie Universitaire; London and St. Louis, B. Herder, 1928.

Conrad von Zabern. *De modo bene cantandi.* Mainz: Peter Schöffer, 1474. Modern edition, *Die Musiktraktate Conrads von Zabern.* Edited by Karl-Werner Gümpel. Akademie der Wissenschaften und der Literatur, Abhandlungen des geistes- und sozialwissenschaftlichen Klasse, Jahrgang 1956, 4. Wiesbaden: F. Steiner for Akademie der Wissenschaften und der Literature, Mainz, 1956.

Crecquillon, Thomas. *Opera omnia.* Edited by Barton Hudson, Mary Tiffany Ferer, and Laura Youens. CMM 63. N.p.: AIM, 1974–.

Crocker, Richard L. "Discant, Counterpoint, and Harmony." *JAMS* 15 (1962): 1–21.

Crook, David. *Orlando di Lasso's Imitation Magnificats for Counter-Reformation Munich.* Princeton: Princeton University Press, 1994.

Cross, Ronald. "Matthaeus Pipelare: A Historical and Stylistic Study of His Works." 2 vols. Ph.D. dissertation, New York University, 1961.

———. "The Life and Works of Matthaeus Pipelare." *Musica Disciplina* 17 (1963): 97–114.

Cummings, Anthony M. "Toward an Interpretation of the Sixteenth-Century Motet." *JAMS* 34 (1981): 43–59.

D'Accone, Frank, ed. *Music of the Florentine Renaissance.* 12 vols. CMM 32. Rome: AIM, 1966–96.

Dean, Jeffrey. "The Evolution of a Canon at the Papal Chapel: The Importance of Old Music in the Fifteenth and Sixteenth Centuries." In Richard Sherr, ed., *Papal Music and Musicians in Late Medieval and Renaissance Rome,* 138–66. Oxford: Clarendon, in association with the Library of Congress, Washington, D. C., 1998.

———. "Listening to Sacred Polyphony *c.* 1500." *Early Music* 25 (1997): 611–36; 26 (1998): 537–39.

———. "Okeghem's Valediction? The Meaning of *Intemerata Dei mater.*" In Philippe Vendrix, ed., *Johannes Ockeghem: Actes du XLe Colloque international d'études humanistes, Tours, 3–8 février 1997,* 521–70. [Paris]: Klincksieck, 1998.

Devaux, Jean. *Jean Molinet: Indiciaire bourguignon.* Bibliothèque du XVe siècle 55. Paris: Honoré Champion, 1996.

Disertori, Benvenuto. *Le frottole per canta e liuto intabulate da F. Bossinensis.* Milan: Ricordi, 1964.

Dobbins, Frank. "Valois." *New Grove II.*

Donato, Baldassare. *Il primo libro de motetti a cinque, a sei, et otto voci novamente composti, & dati in luce (Venice: Gardano, 1599).* Edited by Richard Sherr. The Sixteenth-Century Motet 30. New York and London: Garland, 1994.

Dronke, Peter. *Medieval Latin and the Rise of the European Love-Lyric.* 2 vols. Oxford: Clarendon, 1965–66.

———. "The Song of Songs and Medieval Love-Lyric." In W. Lourdaux and D. Verhelst, eds., *The Bible and Medieval Culture,* 236–62. Mediaevalia Lovaniensia I/7. Leuven: Leuven University Press, 1979.

Dumoulin, Chanoine Jean et al. *La Messe de Tournai: Une messe polyphonique en l'honneur de Notre-Dame à la Cathédrale de Tournai au XIVe siècle.* Tournai: Archives du Chapitre Cathédral, and Louvain-la-Neuve: Université Catholique de Louvain, 1988.

Dunning, Albert. *Die Staatsmotette, 1480–1555.* Utrecht: A. Oosthoek, 1970.

Dupire, Noël. *Jean Molinet: La Vie—Les Œuvres.* Paris: E. Droz, 1932.

Eitner, Robert, ed. *Sechzig Chansons zu vier Stimmen aus der ersten Hälfte des 16. Jahrhunderts von französischen und niederländischen Meistern.* Publikationen älterer praktischer und theoretischer Musikwerke 23. Leipzig: Breitkopf & Härtel, 1899.

Elders, Willem. "New Light on the Dating of Josquin's *Hercules* Mass." *TVNM* 48 (1998): 112–49.

———. "Plainchant in the Motets, Hymns, and Magnificat of Josquin des Prez." In *JosCon,* 523–42.

Elias, Cathy Ann. "Imitation, Fragmentation, and Assimilation of Chansons in the Masses of Gombert, Clemens, and Crecquillon: A Kaleidoscopic Process." Ph.D. dissertation, University of Chicago, 1994.

Fallows, David. *A Catalogue of Polyphonic Songs, 1415–1480.* Oxford: Oxford University Press, 1999.

———. *Dufay.* London: J. M. Dent, 1982.

———. "Embellishment and Urtext in the Fifteenth-Century Song Repertories." *Bäsler Jahrbuch für historische Musikpraxis* 14 (1990): 59–85.

———. "Johannes Ockeghem: The Changing Image, the Songs, and a New Source." *Early Music* 12 (1984): 218–30.

———. "Josquin and Milan." *Plainsong and Medieval Music* 5 (1996): 69–80.

———. "Specific Information on the Ensembles for Composed Polyphony, 1400–1474." In Stanley Boorman, ed., *Studies in the Performance of Late Medieval Music,* 109–59. Cambridge: Cambridge University Press, 1983.

Fassler, Margot. *Gothic Song: Victorine Sequences and Augustinian Reform in Twelfth-Century Paris.* Cambridge: Cambridge University Press, 1993.

Feininger, Laurence, ed. *Missae de G. Dufay et Anonymi.* Monumenta Polyphoniae Liturgicae Sanctae Ecclesiae Romanae I/2. Rome: Societas Universalis Sanctae Cecilae, 1952.

———, ed. *Missa prima [–VI] super L'homme armé, auctore ignoto, secundum unicum fontem mutilem, scilicet Codicem Neapolitanum Ms. VI-E.40.* Monumenta Polyphoniae Liturgicae Sanctae Ecclesiae Romanae I/3. Rome: Societas Universalis Sanctae Cecilae, 1957–74.

Ferand, Ernest T. "Didactic Embellishment Literature in the Late Renaissance: A Survey of Sources." In Jan LaRue et al., eds., *Aspects of Medieval and Renaissance Music: A Birthday Offering to Gustave Reese,* 154–72. New York: W. W. Norton, 1966; reprint ed., New York: Pendragon, 1978.

———. "Improvised Vocal Counterpoint in the Late Renaissance and Early Baroque." *Annales musicologiques* 4 (1956): 129–74.

Ferrand, Françoise. "Le Grand Rhétoriqueur Jean Molinet et la chanson polyphonique à la Cour des Ducs de Bourgogne." In Danielle Buschinger and André Crépin, eds., *Musique, littérature et société au moyen âge: Actes du colloque, 24–29 Mars 1980,* 395–407. Paris: Honoré Champion, for Université de Picardie, Centre d'études médiévales, 1980.

Festa, Costanzo. *Opera omnia.* 8 vols. Edited by Alexander Main and Albert Seay. CMM 25. N.p.: AIM, 1962–78.

Févin, Antoine de. *Les Œuvres complètes/Collected Works.* 4 vols. Edited by Edward Clinkscale. Gesamtausgabe 11. Henryville, Ottawa, and Binningen: Institute of Mediaeval Music, 1980–96.

Finscher, Ludwig. "Loyset Compère and His Works, IV: The Masses and Mass Sections." *Musica Disciplina* 13 (1959): 123–54.

Fischer, Kurt von. "Bemerkungen zur Überlieferung und zum Stil der geistlichen Werke des Antonius dictus Zacharias de Teramo." *Musica Disciplina* 41 (1987): 161–82.

———. "Neue Quellen zum einstimmigen Ordinariumszyklus des 14. und 15. Jahrhunderts aus Italien." In Albert van der Linden, ed., *Liber amicorum Charles van den Borren,* 60–68. Antwerp: Lloyd Anversois, 1964.

———. "Die Rolle der Mehrstimmigkeit am Dome von Siena zu Beginn des 13. Jahrhunderts." *Archiv für Musikwissenschaft* 18 (1961): 167–82.
Fischer, Kurt von, and Max Lütolf, eds. *Handschriften mit mehrstimmiger Musik des 14., 15. und 16. Jahrhunderts.* 2 vols. RISM B IV$^{3-4}$. Munich-Duisburg: G. Henle, 1972.
Fitch, Fabrice. *Johannes Ockeghem: Masses and Models.* Collection Ricercar 2. Paris: Honoré Champion, 1997.
Fleming, John V. *The Roman de la Rose: A Study in Allegory and Iconography.* Princeton: Princeton University Press, 1969.
Flotzinger, Rudolf. "Non-mensural Sacred Polyphony (*Discantus*) in Medieval Austria." In Cesare Corsi and Pierluigi Petrobelli, eds., *Le Polifonie primitive in Friuli e in Europa: Atti del congresso internazionale Cividale del Friuli, 22–24 agosto 1980,* 43–61. Rome: Edizione Torre d'Orfeo, 1989.
Frey, Herman-Walther. "Regesten zur päpstlichen Kapelle unter Leo X. und seiner Privatkapelle." *Musikforschung* 8 (1955): 58–73, 178–99, 412–37; 9 (1956): 46–57, 139–56, 411–19.
Fromson, Michele. "Imitation and Innovation in the North-Italian Motet, 1560–1605." Ph.D. dissertation, University of Pennsylvania, 1988.
———. "The Sixteenth-Century Motet: An Update on Published Catalogues and Indexes in Progress." *Notes* 52 (1995–96): 45–54.
———. "Themes of Exile in Willaert's *Musica nova.*" *JAMS* 48 (1994): 442–87.
Fulton, Rachel. "Mimetic Devotion, Marian Exegesis, and the Historical Sense of the Song of Songs." *Viator: Medieval and Renaissance Studies* 27 (1996): 85–116.
Gabrieli, Andrea. *Edizione nationale delle opere.* Edited by Denis Arnold and David Bryant. Milan: Ricordi, 1988–.
Gallo, F. Alberto. "The Practice of *Cantus Planus Binatim* in Italy from the Beginning of the 14th to the Beginning of the 16th Century." In Cesare Corsi and Pierluigi Petrobelli, eds., *Le Polifonie primitive in Friuli e in Europa: Atti del congresso internazionale Cividale del Friuli, 22–24 agosto 1980,* 13–80. Rome: Edizione Torre d'Orfeo, 1989.
Ganassi, Sylvestro di. *Opera intitulata Fontegara.* Venice: author, 1535.
Gardner, Edmund. *Dukes and Poets in Ferrara: A Study in the Poetry, Religion and Politics of the Fifteenth and Early Sixteenth Centuries.* New York: A. Constable, 1904.
Geering, Arnold. *Die Organa und mehrstimmigen Conductus in den Handschriften des deutschen Sprachgebietes vom 13. bis 16. Jahrhundert.* Bern: P. Haupt, 1952.
Gerber, Rebecca. "Dufay's Style and the Question of Cyclic Unity in the Trent 88 Mass Proper Cycles." In Peter Wright, ed., *I codici musicale trentini: Nuove scoperte e nuovi orientamenti della ricerca,* 107–19. Trent: Provincia autonoma di Trento, Servizi[o] beni librari e archivistici, 1986.
Gerson, Jean. *Early Works.* Translated by Brian Patrick McGuire. Classics of Western Spirituality 92. New York: Paulist, 1998.
———. *Œuvres complètes.* Edited by Palémon Glorieux. Paris and New York: Desclée, 1960–73.
Giller, Don. "The Naples L'Homme Armé Masses and Caron: A Study in Musical Relationships." *Current Musicology* 32 (1981): 7–28.
Glarean, Heinrich. *Dodecachordon.* 2 vols. Translated by Clement A. Miller. Musicological Studies and Documents 6. N.p.: AIM, 1965.
Gold, Penny Schine. *The Lady and the Virgin: Image, Attitude, and Experience in Twelfth-Century France.* Chicago: University of Chicago Press, 1985.
Goldberg, Clemens. "Militat omnis amans: Zitat und Zitieren in Molinets 'Le débat du viel Gendarme et du viel amoureux' und Ockeghems Chanson 'L'autre d'antan.'" *Musikforschung* 42 (1989): 341–49.
Gombert, Nicolas. *Opera omnia.* 11 vols. Edited by Joseph Schmidt-Görg. CMM 6. Rome: AIM, 1951–75.
Gomez Muntané, Maria. "El ars antigua en Cataluña." *Revista de musicologia* 2 (1979): 279–84.
Graduale secundum morem sancte Romane ecclesie. Venice: Antonius de Giunta, 1499.
Guerrero, Francesco. *Ave Virgo sanctissima.* Edited by Bruno Turner. Renaissance Performing Scores A, Spanish Church Music 28. London: Mapa mundi, 1979.
Gümpel, Karl-Werner. "Cantus Eugenianus–Cantus melodicus." In Daniel Heartz and Bonnie Wade, eds., *International Musicological Society: Report of the Twelfth Congress, Berkeley 1977,* 407–13. Kassel, Basel, and London: Bärenreiter, for the American Musicological Society, 1981.
Haar, James. "Some Remarks on the 'Missa La sol fa re mi.'" In *JosCon,* 564–88.

Haggh, Barbara Helen. "Communication." *JAMS* 40 (1987): 139–43.

Harrán, Don. *In Defense of Music: The Case for Music as Argued by a Singer and Scholar of the Late Fifteenth Century*. Lincoln: University of Nebraska Press, 1989.

Higgins, Paula Marie. "Antoine Busnois and Musical Culture in Late Fifteenth-Century France and Burgundy." Ph.D. dissertation, Princeton University, 1987.

———. "Music and Musicians at the Sainte-Chapelle of the Bourges Palace, 1405–1515." In Angelo Pompilio et al., eds., *Atti del XIV Congresso della Società Internazionale di Musicologia: Trasmissione e recezione delle forme di cultura musicale*, 3:689–701. Turin: EDT, 1990.

Hughes, Andrew. *Medieval Manuscripts for Mass and Office: A Guide to Their Organization and Terminology*. Toronto: University of Toronto Press, 1982.

Huizinga, Johan. *Homo Ludens: A Study of the Play-Element in Culture*. Boston: Beacon, 1950.

———. *The Waning of the Middle Ages: A Study of the Forms of Life, Thought and Art in France and the Netherlands in the XIVth and XVth Centuries*. Revised and translated under the author's supervision by F[rederik] Hopman. London: E. Arnold, 1924. Translated from *Herfsttij der Middeleeuwen: Studie over levens- en gedachtenvormen der veertiende en vijftiende eeuw in Frankrijk en de Nederlanden*. Haarlem: H. D. Tjeenk Willink, 1919.

Huot, Sylvia. *Allegorical Play in the Old French Motet: The Sacred and the Profane in Thirteenth-Century Polyphony*. Stanford: Stanford University Press, 1997.

Isaac, Heinrich. *Messen*. 2 vols. Edited by Martin Staehelin. Musikalische Denkmäler 7–8. Mainz: B. Schott, 1970–73.

———. *Opera omnia*. 8 vols. to date. Edited by Edward R. Lerner. CMM 65. N.p.: AIM, 1974–.

Jacquet of Mantua. *Opera omnia*. 6 vols. Edited by Philip T. Jackson and George Nugent. CMM 54. N.p.: AIM, 1971–86.

Jas, Eric. "Nicolas Gombert's *Missa Fors seulement*: A Conflicting Attribution." *Revue belge de musicologie/Belgisch tijdschrift voor muziekwetenschap* 46 (1992): 163–77.

Jeffery, Peter. *Re-Envisioning Past Musical Cultures: Ethnomusicology in the Study of Gregorian Chant*. Chicago: University of Chicago Press, 1992.

Jeppesen, Knud. "Die 3 Gafurius-Kodizes der Fabbrica del Duomo, Milano." *Acta Musicologica* 3 (1931): 14–28.

Jeppesen, Knud, and Viggo Brondal. *Die Mehrstimmige italienische laude um 1500*. Leipzig: Breitkopf & Härtel, 1935.

Josephson, Nors S. "Michot, Andreas." *New Grove II*.

———, ed. *Early Sixteenth-Century Sacred Music from the Papal Chapel*. CMM 95. Neuhausen-Stuttgart: Hänssler/AIM, 1982.

Judd, Cristle Collins. *Reading Renaissance Music Theory: Hearing with the Eyes*. Cambridge: Cambridge University Press, 2000.

Kellman, Herbert. "Josquin and the Courts of the Netherlands and France: The Evidence of the Sources." In *JosCon*, 181–216.

Kemp, Walter H. *Burgundian Court Song in the Time of Binchois: The Anonymous Chansons of El Escorial, MS V.III.24*. Oxford: Clarendon, 1990.

Kirkman, Andrew. "The Invention of the Cyclic Mass." *JAMS* 54 (2001): 1–47.

———. *The Three-Voice Mass in the Later Fifteenth and Early Sixteenth Centuries: Style, Distribution, and Case Studies*. New York: Garland, 1995.

Kirn, Paul. *Friedrich der Weise und die Kirche: seine Kirchenpolitik vor und nach Luthers Hervortreten im Jahre 1517*. Beiträge zur Kulturgeschichte des Mittelalters und der Renaissance 30. Leipzig and Berlin: Teubner, 1926.

Kreider, John Evan. "The Masses for Five and Six Voices by Pierre de la Rue." 2 vols. Ph.D. dissertation, Indiana University, 1974.

Kripke, Saul A. *Wittgenstein on Rules and Private Language: An Elementary Exposition*. Cambridge, Mass.: Harvard University Press, 1982.

Kristeva, Julia. "Bakhtine, le mot, le dialogue et le roman." *Critique* 23 (1967): 438–65.

Lane, Barbara G. *The Altar and the Altarpiece: Sacramental Themes in Early Netherlandish Painting*. New York: Harper and Row, 1984.

Laor, Lia. "Concerning the Liturgical Usage of Dufay's Fragmentary Masses." *Current Musicology* 37/38 (1984): 49–58.

La Rue, Pierre de. *Opera omnia*. 9 vols. to date. Edited by Nigel St. John Davison, J. Evan Kreider, and T. Herman Keahey. CMM 97. Neuhausen-Stuttgart: Hänssler/AIM, 1989–.

Lasso, Orlando di. *Sämtliche Werke*. 21 vols. Edited by Franz Xaver Haberl and Adolf Sandberger. Leipzig: Breitkopf & Härtel, 1894–1926.

———. *The Seven Penitential Psalms and Laudate Dominum de caelis*. Edited by Peter Bergquist. Recent Researches in the Music of the Renaissance 86–87. Madison, Wisconsin: A-R, 1990.

La Trobe University Medieval Music Database. Prepared by John Stinson. http://www.lib.latrobe.edu.au/Audio-Visual/Stinson/medmusic.htm.

Layolle, Francesco de. *Masses and Penitential Psalms*. Edited by Frank D'Aconne. CMM 32/6. N.p.: AIM, 1973.

Leclercq, Jean. *Monks and Love in Twelfth-Century France: Psycho-Historical Essays*. Oxford: Clarendon, 1979.

Leverett, Adelyn Peck. "Song Masses in the Trent Codices: The Austrian Connection." *Early Music History* 14 (1995): 205–56.

Lewis, Mary S., ed. *The Gardane Motet Anthologies*. The Sixteenth-Century Motet 13. New York: Garland, 1993.

Lheritier, Johannes. *Opera omnia*. 2 vols. Edited by Leeman L. Perkins. CMM 48. N.p.: AIM, 1969.

Linden, Albert van der. "La Musique dans les Chroniques de Jean Molinet." In *Mélanges Ernest Closson: Recueil d'articles musicologiques offert à Ernest Closson à l'occasion de son soixante-quinzième anniversaire*, 166–80. Brussels: Société belge de musicologie, 1948.

Llorens, Josephus [José] M. *Capellae Sixtinae Codices musicis notis instructi sive manu scripti sive praelo excussi*. Studi e testi 102. Vatican City: Biblioteca Apostolica Vaticana, 1960.

Lockwood, Lewis. *The Counter-Reformation and the Masses of Vincenzo Ruffo*. Venice: Fondazione Giorgio Cini, [1970].

———. "Introduction." In *The Collected Works of Josquin des Prez*, vol. 11: *Masses Based on Solmisation Themes*, xi–xii. Utrecht: Koninklijke Vereniging voor Nederlandse Muziekgeschiedenis, 2002.

———. "Josquin at Ferrara: New Documents and Letters." In *JosCon*, 103–37.

———. *Music in Renaissance Ferrara, 1400–1505: The Creation of a Musical Center in the Fifteenth Century*. Studies in the History of Music 2. Oxford: Clarendon; New York: Oxford University Press; Cambridge, Mass.: Harvard University Press, 1984.

———. "On 'Parody' as Term and Concept in 16th-Century Music." In Jan La Rue et al., eds., *Aspects of Medieval and Renaissance Music: A Birthday Offering to Gustave Reese*, 560–75. New York: W. W. Norton, 1966; reprint ed., New York: Pendragon, 1978.

———. "A View of the Early Sixteenth-Century Parody Mass." In Albert Mell, ed., *Queens College of the City University of New York, Department of Music: Twenty-Fifth Anniversary Festschrift (1937–1962)*, 53–78. Flushing, N. Y.: Queens College Press, 1964.

Lockwood, Lewis, Noel O'Regan, and Jessie Ann Owens. "Palestrina, Giovanni Pierluigi da." *New Grove II*.

Lombardi, Sandro. *Jean Fouquet*. Florence: Libreria Editrice Salimbeni, 1983.

Long, Michael. "Symbol and Ritual in Josquin's *Missa Di Dadi*." *JAMS* 42 (1989): 1–22.

Lowinsky, Edward E. "Ascanio Sforza's Life: A Key to Josquin's Biography and an Aid to the Chronology of His Works." In *JosCon*, 31–75.

———. "Josquin des Prez and Ascanio Sforza." In Maria Luisa Gatti Perer, ed., *Il duomo di Milano: Congresso internazionale, Milano, Museo della scienza e della tecnica, 8, 12 settembre 1968, Atti*, 2:17–22. Monografie di arte lombarda/I monumenti 3. Milan: La Rete, 1969.

Lowinsky, Edward E., and Bonnie J. Blackburn, eds. *Josquin des Prez: Proceedings of the International Josquin Festival-Conference held at the Juilliard School at Lincoln Center in New York City, 21–25 June 1971*. London, New York, and Toronto: Oxford University Press, 1976.

Ludwig, Friedrich. *Repertorium organorum recentioris et motetorum vetustissimi stili, I: Catalogue raisonné der Quellen*. 2 vols. Vol. 1, Halle: M. Niemeyer, 1910; reprint ed., Brooklyn: Institute of Mediaeval Music, 1964; Vol. 2, n.p.: Institute of Mediaeval Music, 1978.

Lusitano, Vicente. *Introdutione facilissima et novissima*. Rome: Antonio Blado, 1553.

MacClintock, Carol. "Molinet, Music, and Medieval Rhetoric." *Musica Disciplina* 13 (1959): 109–21.

Macey, Patrick. *Bonfire Songs: Savonarola's Musical Legacy*. Oxford: Clarendon, 1998.

———. "Some Thoughts on Josquin's *Illibata dei virgo nutrix* and Galeazzo Maria Sforza." In Albert Clement and Eric Jas, eds., *From Ciconia to Sweelinck: Donum Natalicium Willem Elders*, 111–24. Chloe/Beihefte zum Daphnis 21. Amsterdam and Atlanta: Rodopi, 1994.

Mâle, Émile. *Religious Art in France, The Thirteenth Century: A Study of Medieval Iconography and Its Sources.* Translated by Marthiel Mathews. Bollingen Series 90/2. Princeton: Princeton University Press, 1984. Translated from *L'Art religieux du XIIIe siècle en France: Étude sur l'iconographie du moyen âge et sur ses sources d'inspiration.* Paris: Armand Colin, 1898; 9th ed., 1958.

Marshall, Robert. "The Paraphrase Technique of Palestrina in His Masses Based on Hymns." *JAMS* 16 (1963): 356–58.

Martini, Johannes. *Masses.* Edited by Elaine Moohan and Murray Steib. Recent Researches in the Music of the Middle Ages and Early Renaissance 34–35. Madison: A-R, 1999.

Matter, E. Ann. *The Voice of My Beloved: The Song of Songs in Western Medieval Christianity.* Philadelphia: University of Pennsylvania Press, 1990.

Mattfield, Jacquelyn A. "Some Relationships between Texts and Cantus Firmi in the Liturgical Motets of Josquin des Pres." *JAMS* 14 (1961): 159–83.

McPeek, Gwynn S., and Robert White Linker, eds. *The British Museum Manuscript Egerton 3307.* London: Oxford University Press, 1963.

Meconi, Honey. "Does *Imitatio* Exist?" *Journal of Musicology* 12 (1994): 152–78.

———. "French Print Chansons and Pierre de la Rue: A Case Study in Authenticity." In Jessie Ann Owens and Anthony M. Cummings, eds., *Music in Renaissance Cities and Courts: Studies in Honor of Lewis Lockwood,* 187–214. Warren, Mich.: Harmonie Park Press, 1997.

———. "La Rue, Pierre de." *New Grove II.*

———. "Listening to Sacred Polyphony." *Early Music* 26 (1998): 375–79, 539.

———. "Petrucci's Mass Prints and the Naming of Things." In *Proceedings of the Conference Venezia 1501: Petrucci e la stampa musicale.* Venice: Fondazione Levi, forthcoming.

———. *Pierre de la Rue and Musical Life at the Habsburg-Burgundian Court.* Oxford: Oxford University Press, 2003.

———. "Sacred Tricinia and Basevi 2439." *I Tatti Studies: Essays in the Renaissance* 4 (1991): 151–99.

Melnicki, Margareta. *Das einstimmige Kyrie des lateinischen Mittelalters.* Regensburg: Gustav Bosse, 1954.

Merkley, Paul A., and Lora L. M. Merkley. *Music and Patronage in the Sforza Court.* Studi sulla storia della musica in Lombardia/Collana di testi musicologici 3. Turnhout: Brepols, 1999.

Miyazaki, Haruyo. "New Light on Ockeghem's *Missa 'Mi-mi'.*" *Early Music* 13 (1985): 367–75.

Molinet, Jean. *Le faictz et dictz de Jean Molinet.* 3 vols. Edited by Noël Dupire. Paris: Société des anciens textes français, 1936–39.

Morales, Cristóbal de. *Opera omnia: Motetes I–XXV.* Edited by Higinio Anglès. Monumentos de la musica española 13. [Rome]: Consejo superior de Investigaciones Cientificas, Delagación de Roma, 1953.

Natvig, Mary. "The Magnificat Group of Antoine Busnoys: Aspects of Style and Attribution." In Paula Higgins, ed., *Antoine Busnoys: Method, Meaning, and Context in Late Medieval Music,* 257–76. Oxford: Clarendon, 1999.

Noble, Jeremy. "The Function of Josquin's Motets." *TVNM* 35 (1985): 9–31.

———. "Josquin Desprez." *New Grove.*

Noblitt, Thomas. "Die Datierung der Handschrift Mus. ms. 3154 der Staatsbibliothek München." *Musikforschung* 27 (1974): 36–56.

Ockeghem, Johannes. *Collected Works.* 3 vols. 2nd ed. of vols. 1–2. Edited by Dragan Plamenac and Richard Wexler. American Musicological Society Studies and Documents 1, 3, 7. N.p.: American Musicological Society, 1959–92.

———. *Masses and Mass Sections.* 3 vols. in 12 fasc. Edited by Jaap van Benthem. Utrecht: Koninklijke Vereniging voor Nederlandse Muziekgeschiedenis, 1994–.

Osthoff, Helmut. *Josquin Desprez.* 2 vols. Tutzing: Hans Schneider, 1962–65.

Owens, Jessie Ann. *Composers at Work: The Craft of Musical Composition 1450–1600.* New York and Oxford: Oxford University Press, 1997.

Owens, Jessie Ann, and Richard J. Agee. "La stampa della 'Musica nova' di Willaert." *Rivista italiana di musicologia* 24 (1989): 219–305.

Palestrina, Giovanni Pierluigi da. *Le opere complete.* 35 vols. Edited by Raffaele Casimiri et al. Rome: Istituto Italiano per la storia della musica, Edizione Scalera, 1939–99.

Parris, Arthur. "The Sacred Works of Gilles Binchois." 2 vols. Ph.D. dissertation, Bryn Mawr College, 1965.

Pelikan, Jaroslav. *Mary through the Centuries: Her Place in the History of Culture.* New Haven and London: Yale University Press, 1996.

Perkins, Leeman L. "Communication." *JAMS* 40 (1987): 130–34.

———. "The L'Homme Armé Masses of Busnoys and Okeghem: A Comparison." *Journal of Musicology* 3 (1984): 363–96.

———. "Musical Patronage at the Royal Court of France under Charles VII and Louis XI (1422–83)." *JAMS* 37 (1984): 507–66.

———. *Music in the Age of the Renaissance.* New York and London: W. W. Norton, 1999.

———. "Ockeghem, Jean de." *New Grove II.*

Perkins, Leeman L., and Howard Garey, eds. *The Mellon Chansonnier.* 2 vols. New Haven and London: Yale University Press, 1979.

Pesce, Dolores. "Beyond Glossing: The Old Made New in *Mout me fu grief/Robin m'aime/Portare.*" In Dolores Pesce, ed., *Hearing the Motet: Essays on the Motet of the Middle Ages and Renaissance,* 28–51. New York and Oxford: Oxford University Press, 1997.

Phinot, Dominique. *Opera omnia.* 4 vols. Edited by Janez Höfler and Roger Jacobs. CMM 59. N.p.: AIM, 1972–82.

Picker, Martin. "Divitis, Antonius." *New Grove II.*

———. "The Habsburg Courts in the Netherlands and Austria, 1477–1530." In Iain Fenlon, ed., *The Renaissance: From the 1470s to the End of the 16th Century,* 216–42. London: Macmillan, 1989.

———. "Reflections on Ockeghem and *Mi-Mi.*" In Philippe Vendrix, ed., *Johannes Ockeghem: Actes du XLe Colloque international d'études humanistes, Tours, 3–8 février 1997,* 415–32. [Paris]: Klincksieck, 1998.

Pickford, Cedric E. *The Song of Songs: A Twelfth-Century French Version.* London, New York, and Toronto: Oxford University Press for the University of Hull, 1974.

Pigman, G.W. III. "Versions of Imitation in the Renaissance." *Renaissance Quarterly* 33 (1980): 1–32.

Pipelare, Matthaeus. *Opera omnia.* 3 vols. Edited by Ronald Cross. CMM 34. N.p.: AIM, 1966–67.

Pirro, André. "Leo X and Music." Translated by Gustave Reese. *Musical Quarterly* 21 (1935): 1–16.

Plamenac, Dragan. "La Chanson de L'homme armé et MS. VI. E. 40 de la Bibl. Nationale de Naples." In *Annales de la Fédération archéologique et historique de Belgique, Congrès jubilaire,* 229–30. Bruges, 1925.

Planchart, Alejandro Enrique. "Fifteenth-Century Masses: Notes on Performance and Chronology." *Studi musicali* 10 (1981): 5–29.

———. "Guillaume Du Fay's Benefices and His Relationship to the Court of Burgundy." *Early Music History* 8 (1988): 117–71.

Porta, Costanzo. *Opera omnia.* 25 vols. Edited by Siri Cisilino. Padua: Biblioteca Antoniana, 1964–70.

Pozzi, Giovanni. "Codici, stereotipi, topoi e fonte letterarie." In *Intorno al codice: atti del III Convegno della Associazione italiana di studi semiotici, Pavia, 26–27 settembre, 1975,* 37–76. Pubblicazioni della Facoltà di lettere e filosofia dell'Università di Pavia 22. Florence: La nuova Italia, 1976.

Prizer, William F. "Games of Venus: Secular Vocal Music in the Late Quattrocento and Early Cinquecento." *Journal of Musicology* 9 (1991): 3–56.

———. "Isabella d'Este and Lucrezia Borgia as Patrons of Music: The Frottola at Mantua and Ferrara." *JAMS* 38 (1985): 1–33.

Quereau, Quentin W. "Aspects of Palestrina's Parody Procedure." *Journal of Musicology* 1 (1982): 198–216.

Randall, Michael. *Building Resemblance: Analogical Imagery in the Early French Renaissance.* Baltimore and London: The Johns Hopkins University Press, 1996.

Reaney, Gilbert. "Some Little-Known Sources of Medieval Polyphony in England." *Musica Disciplina* 15 (1961): 15–26.

Reese, Gustave. *Music in the Renaissance.* New York: W. W. Norton, 1954.

———. "The Polyphonic 'Missa de Beata Virgine' as a Genre: The Background of Josquin's Lady Mass." In *JosCon,* 589–98.

Reynolds, Christopher A. "The Counterpoint of Allusion in Fifteenth-Century Masses." *JAMS* 45 (1992): 228–60.

———. *Motives for Allusion: Context and Content in 19th-Century Music.* Cambridge, Mass.: Harvard University Press, 2003.

———. "The Origins of San Pietro B 80 and the Development of a Roman Sacred Repertory." *Early Music History* 1 (1981): 257–304.

———. *Papal Patronage and the Music of St. Peter's, 1380–1513.* Berkeley, Los Angeles, and London: University of California Press, 1995.

Rifkin, Joshua. "Beausseron, Johannes Bonnevin alias." *New Grove II.*

———. "Miracles, Motivicity, and Mannerism: Adrian Willaert's *Videns Dominum flentes sorores Lazari* and Some Aspects of Motet Composition in the 1520s." In Dolores Pesce, ed., *Hearing the Motet: Essays on the Motet of the Middle Ages and Renaissance,* 243–64. New York and Oxford: Oxford University Press, 1997.

Ringbom, Sixten. *Icon to Narrative: The Rise of the Dramatic Close-Up in Fifteenth-Century Devotional Painting.* Acta academiae aboensis Ser. A, 31/2. Åbo: Åbo Akademi, 1965.

Robertson, Anne Walters. "The Mass of Guillaume de Machaut in the Cathedral of Reims." In Thomas Forrest Kelly, ed., *Plainsong in the Age of Polyphony,* 100–39. Cambridge and New York: Cambridge University Press, 1992.

Roediger, Karl Erich. *Die geistlichen Musikhandschriften der Universitäts-Bibliothek Jena.* Claves Jenenses: Veröffentlichen der Universitätsbibliothek Jena 3. Jena: Walter Biedermann/Frommannsche Buchhandlung, 1935; reprint ed., Hildesheim, Zürich, and New York: Georg Olms, 1985.

Roth, Adalbert. "Liturgical (and Paraliturgical) Music in the Papal Chapel towards the End of the Fifteenth Century: A Repertory in Embryo." In Richard Sherr, ed., *Papal Music and Musicians in Late Medieval and Renaissance Rome,* 125–37. Oxford: Clarendon, in association with the Library of Congress, Washington, D. C., 1998.

Sabadino degli Arienti, Giovanni. *Art and Life at the Court of Ercole I d'Este: The "De triumphis religionis" of Giovanni Sabadino degli Arienti.* Edited by Werner L. Gundersheimer. Travaux d'humanisme et Renaissance 127. Geneva: Librairie Droz, 1972.

Schmidt, Hans. *Zum Formelhaften Aufbau byzantinischer Kanones.* Wiesbaden: Breitkopf & Härtel, 1979.

Schrade, Leo. "The Cycle of the Ordinarium Missae." In Higini Anglès et al., eds., *In memoriam Jacques Handschin,* 87–96. Strasbourg: P. H. Heitz, 1962.

———. "A Fourteenth Century Parody Mass." *Acta Musicologica* 27 (1955): 13–39.

———. "News on the Chant Cycle of the *Ordinarium Missae.*" *JAMS* 7 (1955): 66–69.

Schreurs, Eugeen. "Petrus Alamire: Music Calligrapher, Musician, Composer, Spy." In Herbert Kellman, ed., *The Treasury of Petrus Alamire: Music and Art in Flemish Court Manuscripts 1500–1535,* 15–27. Ghent and Amsterdam: Ludion, 1999.

Senfl, Ludwig. *Sämtliche Werke.* 11 vols. Edited by Walter Gerstenberg et al. Wolfenbüttel and Zürich: Möseler, 1937–74.

Sherr, Richard. "*Illibata Dei Virgo Nutrix* and Josquin's Roman Style." *JAMS* 41 (1988): 434–64.

———. "The Membership of the Chapels of Louis XII and Anne de Bretagne in the Years Preceding Their Deaths." *Journal of Musicology* 6 (1988): 60–82.

———. "The Papal Chapel ca. 1492–1513 and Its Polyphonic Sources." Ph.D. dissertation, Princeton University, 1975.

———. *Papal Music Manuscripts in the Late Fifteenth and Early Sixteenth Centuries.* Renaissance Manuscript Studies 5. Neuhausen-Stuttgart: Hänssler/AIM, 1996.

———. "Preface." In Richard Sherr, ed., *Papal Music and Musicians in Late Medieval and Renaissance Rome,* ix–xiii. Oxford: Clarendon, in association with the Library of Congress, Washington, D. C., 1998.

———. "The Singers of the Papal Chapel and Liturgical Ceremonies in the Early Sixteenth Century: Some Documentary Evidence." In P. A. Ramsey, ed., *Rome in the Renaissance: The City and the Myth, Papers of the Thirteenth Annual Conference of the Center for Medieval and Early Renaissance Studies,* 249–64. Medieval & Renaissance Texts and Studies 18. Binghamton, N. Y.: Center for Medieval and Early Renaissance Studies, 1982.

———. "The 'Spanish Nation' in the Papal Chapel, 1492–1521." *Early Music* 20 (1992): 601–9.

———, ed. *Selections from Bologna, Civico museo bibliografico musicale, MS Q19: Rusconi Codex.* 2 vols. The Sixteenth-Century Motet 6. New York: Garland, 1989.

Slim, H. Colin, ed. *A Gift of Madrigals and Motets.* 2 vols. Chicago and London: University of Chicago Press, 1972.

Smijers, Albert, and Arthur Tillman Merritt, eds. *Treize livres de motets parus chez Pierre Attaingnant en 1534 et 1535.* 14 vols. Paris: L'Oiseau-Lyre, 1934–64.

Snow, Robert J. "The Mass-Motet Cycle: A Mid-Fifteenth-Century Experiment." In Gustave Reese and Robert J. Snow, eds., *Essays in Musicology in Honor of Dragan Plamenac on His 70th Birthday*," 301–20. Pittsburgh: University of Pittsburgh Press, 1969.

———, ed. *A New-World Collection of Polyphony for Holy Week and the Salve Service: Guatemala City, Cathedral Archive, Music MS 4*. Monuments of Renaissance Music 9. Chicago and London: University of Chicago Press, 1996.

Sparks, Edgar H. *Cantus Firmus in Mass and Motet 1420–1520*. Berkeley and Los Angeles: University of California Press, 1963; reprint ed., New York: Da Capo, 1975.

Stäblein, Bruno. "Messe." *Die Musik in Geschichte und Gegenwart*.

Staehelin, Martin. *Die Messen Heinrich Isaacs*. 3 vols. Bern and Stuttgart: Paul Haupt, 1977.

Steib, Murray. "A Composer Looks at His Model: Polyphonic Borrowing in Masses from the Late Fifteenth Century." *TVNM* 46 (1996): 5–41.

———. "Imitation and Elaboration: The Use of Borrowed Material in Masses from the Late Fifteenth Century." Ph.D. dissertation, University of Chicago, 1992.

Steiner, Robert. "Von Leo X bis Clemens VII: Zwei Humanisten und ein Puritaner auf dem päpstlichen Thron und ihr Verhältnis zur Kunst." In Victor Ravizza, ed., *Festschrift Arnold Geering*, 185–99. Bern and Stuttgart: Paul Haupt, 1972.

Stevens, John. *Words and Music in the Middle Ages: Song, Narrative, Dance and Drama, 1050–1350*. Cambridge and New York: Cambridge University Press, 1986.

Stevenson, Robert. "Josquin in the Music of Spain and Portugal." In *JosCon*, 217–46.

Stewart, Rebecca. "*. . . Ita desiderat anima mea ad te, Deus* (Ps. 42:1): Johannes Ockeghem, a Most Medieval Musician." *TVNM* 47 (1997): 163–200.

Stone, Anne. "Glimpses of the Unwritten Tradition in Some *Ars subtilior* Works." *Musica Disciplina* 50 (1996): 59–93.

Strohm, Reinhard. "Communication." *JAMS* 40 (1987): 576–79.

———. "Die Missa super 'Nos amis' von Johannes Tinctoris." *Musikforschung* 32 (1979): 34–51.

———. *Music in Late Medieval Bruges*. Oxford: Clarendon, 1985.

———. *The Rise of European Music, 1380–1500*. Cambridge: Cambridge University Press, 1993.

Taruskin, Richard. "Antoine Busnoys and the *L'Homme armé* Tradition." *JAMS* 39 (1986): 255–93.

Thein, Wolfgang. *Musikalischer Satz und Textdarbietung im Werk von Johannes Ockeghem*. Würzburger musikhistorische Beiträge 13. Tutzing: Hans Schneider, 1992.

Tinctoris, Johannes. *The Art of Counterpoint (Liber de arte contrapuncti)*. Translated by Albert Seay. Musicological Studies and Documents 5. N.p.: AIM, 1961.

———. *Opera theoretica*. Edited by Albert Seay. Corpus scriptorum de musica 22. N.p.: AIM, 1975–.

———. *Proportions in Music*. Translated by Albert Seay. Colorado Springs: Colorado College Music Press, 1979.

Torchi, Luigi, ed. *L'Arte musicale in Italia*. 7 vols. Milan: Ricordi, n.d.

Treitler, Leo. "'Centonate' Chant: *Übles Flickwerk* or *E pluribus unus*?" *JAMS* 28 (1975): 1–23.

———. "Homer and Gregory: The Transmission of Epic Poetry and Plainchant." *Musical Quarterly* 60 (1974): 333–72.

———. "Music Analysis in an Historical Context." *College Music Symposium* 6 (1966): 75–88.

Tuohy, Thomas. *Herculean Ferrara: Ercole d'Este, 1471–1505, and the Invention of a Ducal Capital*. Cambridge and New York: Cambridge University Press, 1996.

Ulff-Møller, Nina K. "The Connection between Melodic Formulas and Stereotype Text Phrases in Slavonic Stichera." *Cahiers de l'Institut du Moyen-âge grec et latin, Université de Copenhague* 54 (1986): 49–60.

Van Doorslaer, G[eorges]. "La Chapelle musicale de Philippe le Beau." *Revue belge d'archéologie et d'histoire de l'art* 4 (1934): 21–57, 139–65.

Vaughan, Richard. *Charles the Bold: The Last Valois Duke of Burgundy*. London: Longman, 1973.

———. *Philip the Good: The Apogee of Burgundy*. Harlow: Longmans, 1970.

———. *Valois Burgundy*. London: Penguin, 1975.

Victoria, Tomás Luis de. *Opera omnia*. 8 vols. Edited by Felipe Pedrell. Leipzig: Breitkopf & Härtel, 1902–13.

Wagner, Peter. *Geschichte der Messe*. Kleine Handbücher der Musikgeschichte nach Gattungen 11/1. Leipzig: Breitkopf & Härtel, 1913.

Warmington, Flynn. "A Survey of Scribal Hands in the Manuscripts." In Herbert Kellman, ed., *The Treasury of Petrus Alamire: Music and Art in Flemish Court Manuscripts 1500–1535*, 41–46. Ghent and Amsterdam: Ludion, 1999.

Warner, Marina. *Alone of All Her Sex: The Myth and Cult of the Virgin Mary.* New York: Knopf, 1976.
Weber, Edith. *Le Concile de Trente et la musique: De la Réforme à la Contre-Réforme.* Paris: Honoré Champion, 1982.
Wegman, Rob C. "The Anonymous Mass *D'Ung aultre amer*: A Late Fifteenth-Century Experiment." *Musical Quarterly* 74 (1990): 566–94.
———. "Another 'Imitation' of Busnoys's *Missa L'Homme armé*–and Some Observations on *Imitatio* in Renaissance Music." *JRMA* 114 (1989): 189–202.
——— "Another Mass by Busnoys?" *Music and Letters* 71 (1990). 1–19.
———. *Born for the Muses: The Life and Masses of Jacob Obrecht.* Oxford: Clarendon, 1994.
———. "From Maker to Composer: Improvisation and Musical Authorship in the Low Countries, 1450–1500." *JAMS* 49 (1996): 409–79.
———. "*Miserere supplicanti Dufay*: The Creation and Transmission of Guillaume Dufay's *Missa Ave regina celorum*." *Journal of Musicology* 13 (1995): 18–54.
Wilkins, Nigel E., ed. *A 15th-Century Repertory from the Codex Reina (Paris: Bibl. nat., nouv. acq. Fr. 6771).* CMM 37. [Rome]: AIM, 1966.
Willaert, Adrian. *Opera omnia.* Edited by Hermann Zenck et al. CMM 3. Rome: AIM, 1950–.
Wright, Craig. "Fontaine, Pierre." *New Grove II.*
———. *Music and Ceremony at Notre Dame of Paris 500–1550.* Cambridge: Cambridge University Press, 1989.
———. "Performance Practices at the Cathedral of Cambrai 1475–1550." *Musical Quarterly* 64 (1978): 295–328.
Zarlino, Gioseffo. *The Art of Counterpoint: Part Three of Le istitutioni harmoniche, 1558.* Translated by Guy A. Marco and Claude V. Palisca. New Haven and London: Yale University Press, 1968.
———. *On the Modes: Part Four of Le Istitutioni harmoniche, 1558.* Translated by Vered Cohen. New Haven and London: Yale University Press, 1983.
Ziino, Agostino. "Polifonia nella cattedrale di Lucca durante il XIII secolo." *Acta musicologica* 47 (1975): 16–30.
Zumthor, Paul. *Langue et techniques poétiques à l'époque romaine (XIe–XIIIe siècles).* Paris: C. Klincksieck, 1963.
———. *Le Masque et la lumière: La Poétique des grands rhétoriqueurs.* Paris: Seuil, 1978.
———. "Style and Expressive Register in Medieval Poetry." In Seymour Chatman, ed. and (in part) trans., *Literary Style: A Symposium*, 263–91. London and New York: Oxford University Press, 1971.

Contributors

M. Jennifer Bloxam is Professor of Music at Williams College. Her writings on the use of preexistent material in late medieval masses and motets have appeared in *JAMS, Early Music History, Journal of Musicology,* the *Josquin Companion,* and other publications. Work in progress includes a book on the ritual and theological context of late medieval masses based on multiple plainsong cantus firmi.

Cathy Ann Elias is Assistant Professor of Musicianship Studies at DePaul University. Her research interests include compositional process and performance practice. Her most recent article, "Sercambi's *Novelliere* and *Croniche* as Evidence for Musical Entertainment in the Fourteenth Century," is forthcoming in *The Italian Novella* (Routledge, 2002).

Michele Fromson has written extensively about the Counter-Reformation motet, late-sixteenth-century music theory, and Adrian Willaert, with her work appearing in *JAMS, JRMA, Notes, Theory and Practice,* and *New Grove II.* She has taught at Columbia University and other institutions, and is currently Associate Director of Educational Outreach and Development for the San Francisco Contemporary Music Players.

Jenny Hodgson is an independent scholar working in Lancashire, England, who has studied with Stephen Rees, Rob Wegman, and David Fallows. Her writings appear in *Early Music,* and her edition of the works of Firmin Caron will be published by the American Institute of Musicology.

Honey Meconi is author of *Pierre de la Rue and Musical Life at the Habsburg-Burgundian Court* and numerous shorter publications, as well as editor of *Fortuna desperata: Thirty-Six Settings of an Italian Song.* She is director of Medieval Studies at Rice University and is currently writing a book on the music of Hildegard von Bingen.

Christopher Reynolds is Professor of Music at the University of California, Davis, and has been Visiting Professor at the University of Heidelberg, Yale University, and the University of California, Berkeley. His most recent book, *Motives for Allusion: Context and Content in 19th-Century Music,* was published by Harvard University Press in 2003.

Murray Steib is Assistant Professor of Music at Ball State University. Editor of *The Reader's Guide to Music: History, Theory, Criticism* and coeditor of Johannes Martini's *Masses,* he has published in *Journal of Musicology, TVNM, New Grove II,* and other venues. He is currently working on a manuscript study, a biography of Johannes Martini, and an edition of his motets.

Andrew H. Weaver is Visiting Assistant Professor of Music at the University of Notre Dame. He received his Ph.D. with distinction from Yale University in 2002, with a dissertation on motets at the Habsburg court in Vienna during the reign of Ferdinand III (1637–57). He is the recipient of both the Hewitt-Oberdoerffer Award and the Hollace Anne Schafer Award. His current projects include an examination of Monteverdi's *Selva morale et spirituale* and an edition of motets by Giovanni Felice Sances.

Index